A Handbook on

Principles and Preparation of Jelly, Jam and Marmalade

THE AUTHOR

Prof. Bibhas Chandra Mazumdar had been in the profession of teaching of post-graduate and under-graduate students of several Indian Universities for more than 34 years and supervised researches towards 72 these of post-graduate students and doctoral fellows. He had also served in fruit development, extension education, farm management and in research positions for more than 7 years. Singly or jointly with others, he has authored 14 books, 163 research papers and 54 non-research articles in horticulture, human nutrition, medicinal plants and plant physiology.

A Handbook on Principles and Preparation of Jelly, Jam and Marmalade

– Author –

Dr. Bibhas Chandra Mazumdar
M.Sc. Agri., Ph.D.

Ex-Professor in Horticulture,
Institute of Agricultural Sciences,
Calcutta University;
Reader in Horticulture,
Bidhan Chandra Krishi Vishwavidyalaya, Kalyani;
Lecturer in Horticulture,
Allhabad Agricultural Institute, Allahabad;
Fruit Development Officer,
Burdwan, Govt. of West Bengal.

2024

Daya Publishing House®
A Division of
Astral International Pvt. Ltd.
New Delhi – 110002

ISBN: 978-93-5461-945-8

Published by : **Daya Publishing House®**
A Division of
Astral International Pvt. Ltd.
– ISO 9001:2015 Certified Company –
4736/23, Ansari Road, Darya Ganj
New Delhi-110 002
Ph. 011-43549197, 23278134
e-mail: info@astralint.com
Website: www.astralint.com

Printed at : **Replika Press Pvt Ltd**

Dedicated to the venerated memory of

Late T.A. Russell

Author gratefully acknowledges the kind affection and research guidance he received from Late T.A. Russel, the magnanimous British horticulturist at the then Allahabad Agricultural Institute in the early part of 1971 when he has served as a Lecturer in Horticulture in the same Institute.

Preface

Necessity is the mother of invention and perishability of fruits in fresh form is responsible for giving birth to the technology of processing and preservation of fruits. Nicolas Appert deserves credit for this novel area of food science and technology. Despite the fact that preserved fruits could not be regarded to be equipotential to fresh fruits in respect of health giving properties, their necessity cannot be set at defiance under certain conditions. Barring their supply in hotels, restaurants and use of tourists, these items have a dire necessity to the defence personnels during war. Fruits having less orgnoleptic appeal also find utilization in making many delicious products.

Plenteous products are in fact, made from fruits and among these, jelly, jam and marmalade are considered of maximum demand across the world. Popularity of these products are known to be very high in France, Germany and other European countries. Production of jelly, jam and marmalade is now considerably high in India, in which country, output is estimated to be about 54 thousand metric tonnes. The country grows hundreds of species of fruits in established plantations and these apart, hundreds of species grow as wild. Evidently, fruits of many wild and less-known species could be successfully exploited in making jelly, jam and marmalade. Vast potentiality exists for large-scale expansion of jelly, jam and marmalade production in small and household scale in the country without much capital investment and the avocation is lucrative.

The author acknowledges Late Suman Chandra Mallik, the then Professor of Horticulture at the B.C.K.V. Kalyani, West Bengal in gathering from him valuable knowledge in this technology. Grateful acknowledgement is due to many other persons, among which, Late Bechu, the laboratory attendant, Horticulture Department of the erstwhile Allahabad Agricultural Institute deserves special mentioning for imparting his self-acquired, practical knowledge to the author while he was a Lecturer in Horticulture in that Institute. The assistance received from Mrs. Purabi Mazumdar, M.A., B.Ed. in various ways in preparation of this book is beyond doubt, unforgettable to the author. Lastly, the author invites valued suggestions from readers of this handbook which has aimed to cater to the requirement of students and personnels having desire to make jelly, jam and marmalade in small and household scale.

Bibhas Chandra Mazumdar

Contents

Chapter 1

Introduction

Stored, preserved or processed fruits are not equipotential to fresh fruits in respect of taste, food value, medicinal properties and other qualities. Hence, instead of taking fruits in fresh form, switching them to storage or preservation is not supported either by the fruit scientists or the food scientists. Nevertheless, necessity of storage and processing to make products cannot be disregarded, at least under certain conditions, among which the following are of pertinence.

Most of the human consumable fruits are seasonally available such that they are not available in all seasons. Storage and preservation are means to taste them out of seasons. Fruits of many species are thin-skinned and highly perishable, *e.g.*, loquat, strawberry, phalsa, cashew-apple, mulberry, fig., some berries *etc.* Such fruits may face difficulty to transport to distant places for marketing in fresh form. Fruits of some species are difficult or perhaps impossible to consume in fresh form, *e.g.*, karonda, otaheite gooseberry, Indian hog-plum, dillenia *etc.* However, these fruits could be utilized to make many delicious preserved products.

The basic aspects of storage and preservation of fruits have been brought to light under the following Heads.

1. Spoilage of Fruits and their Causes

It is a common observation that grains of rice, wheat, maize, pulses, oilseeds *etc.* do not rot when stored for long period under ordinary atmospheric condition. Nuts of almond, cashew, walnut *etc.* also do not rot in long term storage under ordinary atmosphere. On the other hand, ripe fruits of mango, papaya, banana, guava, pineapple *etc.* cannot be stored for long period under ordinary condition. Some fruits like apple, pear, citruses, grapes *etc.* are found to store well for relatively long period but can be stored upto a certain period only.

The pertinent cause may be attributed as the amount of moisture that they contain. The grains stated above do not contain more than 10-13 per cent of water. Water content of nuts (kernels) also does not exceed 4-6 per cent. But in most of the pulpy fruits, water is present from 70-90 per cent. The justification in favour of the fact that water content of fruits is a causative factor for their spoilage during long term storage is evident from the fact that drying or partial removal of water from some fruits which is a common practice in date, fig, grape *etc.* markedly enhance their storage life. However, apart from high moisture, some chemical compounds present in the fruits are also responsible in shortening their storage life under ordinary atmosphere.

Causative Agents

The two pertinent causes of spoilage of fruits are, (a) Microbial and (b) Chemical.

(a) Microbial Spoilage

Spoilage of this type is very often caused by the infection by bacteria or fungi. A number of micro-orgnisms in fact, infect fruits and the common ones may be enlisted as follows:

(i) Bacteria

Bacillus subtilis, B. botulinus, B. cyanogenes, B. viscosus, Streptococcus liquefaciens, S. lactis, Clostridium botulinum, Lactobacillus bulgaricus, Staphylococcus aureus, Escherichia coli, Mycoderma aciti etc.

(ii) Yeast

Saccharomyces ellipsoideus, S. cerevisiae, S. anamalus, S. malei, S. pasleurianus, some Schizosaccharomycetes species *etc.*

(iii) Mould

Species under *Mucor, Aspergillus, Penicillium, Monilia, Botrytis, Dematium Alternaria etc.*

(b) Chemical Spoilage

During ripening of fruits, the autolytic enzymes present in them tend to become active. These enzymes act on the substrates which are present in the fruits and catabolize them, *i.e.*, break down them sequentially. For example, the carbohydrate molecules are broken down into hexose sugars, alcohols, organic acids *etc.* with emission of carbon di-oxide and other gases. Similarly, polysaccharide forms of carbohydrates such as starch is broken down and gives rise to amylose and amylopectin. Cellulose is broken down to hexose surgars and in the same way, pectin is broken down to alpha-D-galacturonic acid molecules. Protein which is the highly complex nitrogenous compound is broken down by the protein-breaking (= proteolytic) enzymes to simpler forms such as proteans, proteoses, peptones, poly-peptides, amino acids and ammonia. The enzymatic breakdown of the complex organic compounds into the simpler forms in the fruits in sequence brings about chemical spoilage in them including emanation of gases which emit offensive smell. The simpler forms of compounds that are formed from the complex compounds also act as substrates of micro-organisms to grow over them.

2. Prevention of Spoilage of Fruits

Spoilage of fruits could be prevented or retarded by two ways. These are,

(a) Storage and

(b) Processing and preservation.

Difference between Storage and Preservation

Both storage and preservation of fruits aim at keeping them over a considerable period of time and thereby, to render them consumable at later time.

(a) Storage

Storage of fruits refer to maintain them without cutting any part, *i.e.*, in intact condition. Any small mechanical injury on the surface of any fruit is however, cut off.

The storage life, *i.e.*, shelf-life of fruits would depend on a number of factors. These are, species and varieties of fruits, their maturity condition, agro-inputs received in their cropping, freedom from the attack by pests, diseases and physiological disorders, harvesting method which has been employed, mode of handling, freedom from injury and above all, the treatments used and the storage atmosphere. Whatever may be the condition, fruits cannot be stored in good condition for longer time.

(b) Processing and Preservation

In processing and preservation, the fruits are cut into pieces or otherwise mechanically injured and preserved by subjecting them to suitable treatments. Therefore, original size and shape of the fruits are not retained when they are processed and preserved.

There are however, exceptions, *e.g.*, *morabba* or preserve of some fruits like aonla, canning of wax apple, litchi, rose-apple, ber *etc.*, in which cases, size and shape of fruits are retained.

The treatments that are used in processing and preservation may be of various types, such as exclusion of moisture to different degrees, addition of agents to prevent microbial infection and physiological rotting and impart taste.

3. Methods of Storage

There are a number of methods of storage of fruits that are followed. The common ones be listed as follows:

(a) Pre-Cooling

This method is followed though rarely when the fruits are required to be transported to distant places and under ordinary atmospheric condition. The fruits are exposed to very low temperature and usually below 0°C. They are maintained at this temperature for 1-5 hours depending on fruits and then brought outside. By such exposure, they become very hard like stone and give metallic sound when they fall on the ground. These pre-cooled fruits require several hours, *i.e.*, 3-20 hours to come to room temperature. During this period, *i.e.*, just after pre-cooling, they are transported. The method is not suitable for all fruits and have a great drawback in that the protoplasms of the cells of fruits form ice and rupture them. The texture may become very soft as well.

(b) Cold Storage

In this method, the fruits are stored in low temperature and usually above 0°C. This refrigeration system maintains temperature by Vapour Absorption or Vapour Compression system.

The actual temperature that is to be maintained in the storage chamber differs according to species of fruits and may differ even according to varieties. The optimum requirement of low temperature and storage duration in that temperature for some species of fruits have been given in Table 1.

Table 1. Low Temperature with Relative Humidity of 85-90 per cent and Storage Duration of some Fruits

Fruit	*Storage*	
	Temperature (°C)	*Duration (Week)*
Mango	7-9	4-7
Banana	11-13	3
Papaya	8-10	1-2
Pineapple	7-9	4-6
Grape	0-2	6-8
Sweet orange	4-7	12-14
Mandarin orange	6-7	10-12
Grapefruit	7-9	8-10
Apple	0-2	15-25
Pear	0-2	12-18
Peach	0-2	2-4
Plum	0-2	4-8

Low temperature storage has a drawback in that, much water is lost from the fruits. To avoid that, it is necessary to maintain high relative humidity in the storage chamber and this should be 80-90 per cent. Maintenance of specified temperature is also very important in the storage chamber, Because, if temperature is increased, autolytic enzymes present in the fruits may result in chemical spoilage of fruits. Again, much lowering of temperature in the cold storage room may cause brownish discoloration in the pulp of some fruits. Very low temperature may also cause protoplasmic rupture of the cells in some fruits.

(c) Controlled Atmospheric Storage

Atmosphere in the storage chamber is modified in this storage. This is done in low temperature storage or without modifying the temperature. Environment in the storage chamber is modified by pumping out some air, moisture increasing carbon di-oxide concentration, occasional removal of ethylene gas, dusting with lime *etc.* Regulation of internal air (1-1.5 cft. per min. per letu) is also maintained.

(d) Gas Storage

This is a controlled atmospheric storage in which gas atmosphere in the storage chamber is only modified without other treatment. Sometimes high concentration of carbon di-oxide is pumped into the storage chamber so that, respiratory activity of the fruits is reduced. Due to high CO_2 concentration, oxygen comes to low level and respiratory rate of fruits is low, eventuating low respiratory breakdown. Microbial growth is also low at high CO_2 concentration. Storage after exposure to sulphur di-oxide gas is also done by exposing over burning sulphur.

(e) Irradiation Storage

Ultra-violet and other forms of ionizing radiations have activity to destroy micro-organisms. Taking that advantage, fruits are exposed to these rays and stored. It has been recently observed that shelf-life of litchis have stretched from 7 to 24 days using gamma radiation showers.

(f) Wax Storage

Plant wax which remains deposited as thin film on fruits or other plant parts is used in this storage. This wax which contains esters of higher fatty acids, alkanes, primary alcohols *etc.* is usually obtained from sugarcane trash (in sugarcane stem, it is abundant) and heaped together as powder. Fruits are dipped in this powder (which is safe) and removed such that a thin layer adheres and then stored.

(g) Chemical Storage

Treatment with some chemical preservatives, *e.g.*, esters of benzoic acid, sorbate, sulphurous acid esters *etc.* are used but in very low concentrations. Nycin, tylosine *etc.* are also sometimes used. Formaldehyde, salicylic acid *etc.* are not used now for their high toxicity. On treatments with chemicals, the fruits are usually stored in relatively low temperature condition.

(h) Domestic Storage

In household scale, a number of methods are followed in storage of fruits. Dipping fruits in weak emulsion of lime water followed by storage after surface-drying is common which slows down ripening by retarding the action of ethylene. Dusting fruits with mixtures of ground spices and condiments is sometimes done. Treatment with common salt is done to store many fruits.

4. Methods of Processing and Preservation

Processing and preservation of fruits is carried on in a number of ways. These depend on type of fruits, their textural condition, chemical components, demand in the markets, infrastructural facilities that are available, cost of production and other factors.

Juicy fruits notably, citruses, pomegranate, phalsa, grape, strawberry, jamun, water-melon *etc.* are most commonly used in making beverages. Among beverages, squashes and crushes are most common. Apart from these, juices, cordial, vinegar are also made from them. Sometimes partially fermented drinks from the juices of such fruits also gets preference.

Apart from juicy fruits, fruits of some species having juicy pulp, *e.g.*, Raspunia variety of mango (sucking varieties) are also used to make beverage. Highly ripe fruits of some species, *e.g.*, papaya, guava, mango, apricot, plum, sapota, bael, banana, peach *etc.* are also known to be used in making drinks or paste.

In the making of jelly and jam evidently, such fruits which are rich sources of pectin of high methoxyl content as well as acidity are preferred and common examples are guava, pear, crab apple, quince, relatives of guava, karonda, wild citruses, cranberry *etc.* In making marmalade, citrus fruits are exclusively used and such citruses which have little thickness of albedo (= mesocarp) and much flavor of flavedo (= epicarp) in their oil-glands are preferred.

For the purpose of making preserve (= *morabba*), aonla, bael, pineapple, pear, peach, apple *etc.* are more commonly used. Fruits of guava having high pectin are used in making fruit-cheese. To make dried fruits, those which are sweet, pulpy with thick skin are usually preferred, *e.g.*, peach, grapes, pear, mango, plum, jackfruit, ber, bael, water-chestnut *etc.* Candy is usually made from apple and pear. Vinegar, which is usually a water solution of acetic acid is a product which is sometimes made from apple (cider vinegar), grape (wine vinegar), orange *etc.*, pickle is usually made by utilizing fruits which are much sour in taste. Canning is a very common method of preserving fruit by storing them in strong sugar solution contained in cans. Quite a number of fruits are used, *e.g.*, apple, mango, grape, mandarin and sweet orange, strawberry, litchi, plum, pear, apricot *etc.* There is also varietal preference of fruits in canning, *e.g.*, in apple, Yellow Newton, Wine sap, Baldwin *etc.*, in mango, alphonso, in grapes, Thompson seedless, in pear, Bartlett, Elberta *etc.*, in strawberry, Talisman, in litchi, Purbi varieties etc. preferred in canning.

It is logical to opine that necessity has been there for fresh fruits, stored fruits as well as processed and preserved fruits. Each of these has merit under particular condition.

In the making of processed fruit products, a scrutiny is necessary which has been discussed in the concluding part.

Chapter 2

Pectic Substances and Gel Setting

The basic chemical ingredient in formation of jelly, jam and marmalade is *pectin.* Hence, some knowledge of this substance has its necessity before actually going to prepare these fruit products. The subject is discussed under the following Heads.

1. Chemical forms.
2. Source of existence.
3. Setting of gel.
4. Jelly-grade.

1. Chemical Forms

The word, pectin is derived from the Greek word, *pepticos* or *pektos*, which means something that makes coagulation, *i.e.*, congelation of some substance particularly in cold condition. This is a chemical compound and apart from pectin, there are also three other chemical compounds which are closely related to it. Collectively, these are called pectic substances and according to the American Chemical Society, these have been termed as follows:

(i) Protopectin,

(ii) Pectinic acid,

(iii) Pectin, and

(iv) Pectic acid

Chemically, the above compounds come under the group of carbohydrates. Carbohydrate compounds are conventionally brought under three broad groups which are termed as monosaccharides, oligosaccharides and polysaccharides. In a monosaccharide molecule, only one sugar unit is present and in an oligosaccharide molecule, 2-10 units are present while the number of units are more than 10 and may be thousands in a polysaccharide molecule. All the four pectic compounds are polysaccharides and these are grouped as *complex polysaccharides.*

Each pectic compound is formed from a type of acid, which is known as D-galacturonic acid. These acid molecules join (= combine) together to form a chain and thereby, a pectic compound is formed. Each individual galacturonic acid molecule is thus, a unit and is considered a monomer. When these monomers combine (= link) together and form the chain, this is called a polymer. It should be noted that the number of monomer units, *i.e.,* galacturonic acid molecules are plentiful in a pectic substance.

Many of these monomers are joined with methyl groups and thus, form methyl esters while some are joined with calcium or magnesium ions and form calcium or magnesium esters.

Pectic compounds (Figure 1) many therefore, be stated as polymers of galacturonic acid esters (= galacturonates) or polygalacturonates.

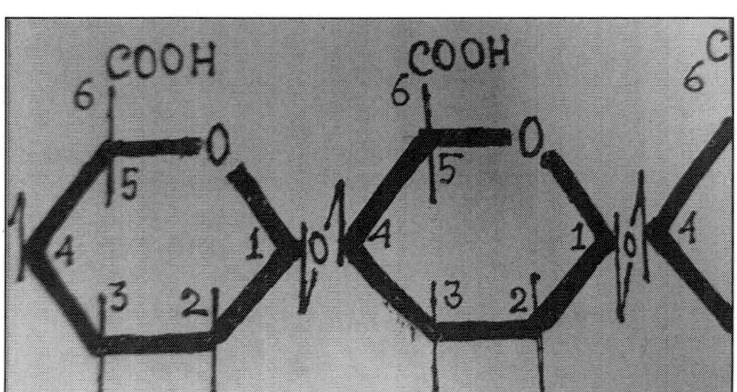

Figure 1. Chemical Structure of Pectic Compound (= alpha-polygalacturonic acid).

It should be noted that among all the four pectic compounds, the length of chain is longest in protopectin. This signifies that the galacturonate monomers are maximum in number in the *protopectin.*

Protopectin does not contain any methyl group and it is not soluble in water. By the action of hydrolyzing enzyme, it forms (= catabolizes) into *pectinic acid.* Length of the chain of this compound becomes shorter than that of the precursor compound, *i.e.,* protopectin. This compound contains some methyl groups and is partly soluble in water. Pectinic acid is again catabolized by enzyme and gives rise to the compound, pectin.

The chain length of pectin becomes still shorter than the compound, pectinic acid. Many monomers of the pectin chain form methyl esters and this compound becomes more soluble in water than the pectinic acid. Pectin is soluble in water to about 20 per cent. Lastly, the compound is enzymatically hydrolyzed, *i.e.,* catabolized to give rise to *pectic acid.* The chain length of the compound, pectic acid is shortest of all the four pectic compounds. It is also highly soluble in water and is most soluble of all the pectic compounds.

Thus, the pectic compounds are sequentially reduced in their chain length by the action of hydrolyzing (= pectolyzing) enzymes and concomitantly, their rate of solubility in water is also increased.

The most important part is, among all the four forms of pectic compounds, *it is pectin which is the most important* of all pectic substances in the formation of jelly, jam or marmalade. This is because, methyl esters are in maximum level in the pectin chain and therefore, maximum amount of the products could be prepared from this compound. (In the International Numbering Systems, pectin has been given a number of 440). However, it should be noted that some amount of jelly, jam or marmalade may be formed from pectinic acid also as it contains some methyl esters although the amount which is formed is very small. Formation of gel, *i.e.,* jelly, jam or marmalade however, does not take place from the compound, protopectin or from the compound, pectic acid.

It is brought to notice in this context that grouping of pectic compounds is also done in consideration to their solubility in water. Accordingly, these are grouped as follows:

a. Water-soluble

Pectic substances, which are soluble in water and these include pectin and colloidal pectinic acids of high methyl ester content.

b. Oxalate-soluble

Pectic substances which include pectic acid and colloidal pectinic acids of low methyl ester content.

c. Acid-soluble and Alkali-soluble

Pectic substances, which include water-soluble forms that have high methyl ester content.

2. Source of Existence

The pectic compounds are present only in the body of plants and never in any animal. But in the animal body, a compound which is similar to pectic compound is present and this is known as gelatin (= gelatine). Although jelly like material is possible to be made from gelatin but its quality is not at par with that prepared from pectin. Hence, in commerce and household scale, it is pectin with which jelly, jam and marmalade are made.

A plant or an animal body is made up of cells. The cells are in fact, the structural and also functional units of the body. The cells of plants and animals have similarity in structure but there are some differences. The main difference is, each plant cell except the reproductive cells has a layer at the boundary and this is known as cell-wall. This wall separates a cell from the surrounding cells and also acts in giving protection to the cells. While protoplasm is the living part of a plant cell, its surrounding structure, *i.e.*, cell-wall is the non-living part of the plant cell.

With maturation of the cell, three distinct layers are formed in the cell-wall. These layers are recognized as (i) middle-lamella, (ii) primary cell-wall and (iii) secondary cell-wall. At first, the middle-lamella is formed and after that, the primary cell-wall and lastly, the secondary cell-wall is formed. The middle-lamella is present at the outer side of the cell-wall. This layer of the cell-wall is thick and its role is to hold the cells firmly together so that they do not loosen apart. The primary cell-wall is thin, elastic and extensible. The secondary cell-wall is thicker than the primary cell-wall but it is thinner than the middle –lamella. The cell-wall and its three layers have been shown in Figure 2.

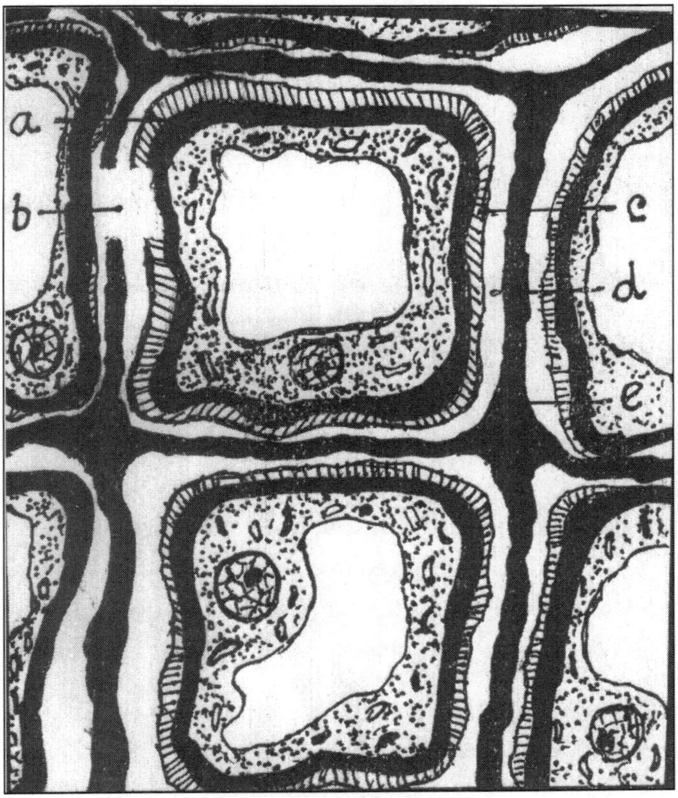

Figure 2. Cell-wall and its Layers.
(a) Cell; (b) Cell-wall; (c) Secondary cell-wall;
(d) Primary cell-wall; (e) Middle-lamella.

A number of compounds are deposited in the cell-wall. These are mostly polysaccharides such as mucilage, cellulose, hemicellulose, lignin, gum *etc.* and among other important polysaccharide compounds that get deposited are pectic substances. Salts, minerals *etc.* are also deposited in the cell-wall. All the materials that are present in the cell-wall are actually produced in the protoplasm and then they are migrated to the cell-wall.

It should be noted that the different ingredients that are present in the cell-wall are not exactly same for all plant species. For example, in the cell-walls of some plant species cellulose is deposited in fairly high amount, in some plant species, mucilage deposition is high and in cell-wall of some species of plants, pectic substances are deposited in much abundance.

It needs mentioning in this regard that the content of the cell-wall also differs according to organs of the plants. For example, in majority of fruits, pectic substances are rich in the cell-walls. Again, it is also noticed that in many fruits, the pectic substances are in relatively high amount at the outer part than at the inner side. As pectin serves the major or principal ingredient in the preparation of jelly, jam and marmalade, evidently, fruits or other materials which are rich sources of pectin, are used in preparation of these fruit products.

3. Setting of Gel

When fruit pieces (or other pectin rich material) from which jelly, jam or marmalade is to be made is suitably boiled in water, the pectin and the other pectic compounds as well as other substances that are water soluble come out in water. After boiling when it is strained, these materials along with pectin are naturally obtained in the strained extract. The extract looks like a *colloidal* solution and not a true solution. (Instead of fruit extract, if synthetic pectin is used to make artificial jelly, then also the solution will be colloidal).

Now, in the colloidal solution which contains pectin, if any dehydrating substance (= which dehydrates, *i.e.*, removes water) is added, the pectin is precipitated (= coagulated). This is called gel setting, *i.e.*, formation of jelly. (The term, gel was proposed by Thomas Graham to denote cross-linked network of the product). In this connection, it may be mentioned that there are many types of dehydrating agents (= water removing substances) which are able to set (= form) gel. Some of these may be named as glycerol, alcohol, sulphuric acid *etc.* However, in order to make jelly, jam or marmalade which are used for consumption by human being, only sugar, *i.e.*, common sugar (= table sugar) is used. This is crude form of the chemical, sucrose. In commerce, this is known as cane sugar (derived from sugarcane, sugarbeet *etc.*) The gel setting will be excellent if the chemical, sucrose is used but this is not used in making jelly because of very high cost.

It is opined by many jelly experts that gel setting becomes perfect if it is done in acidic medium. This may be done by suitable addition of little quantity of organic acid which is not harmful to the human body and preferably, citric acid, or by addition of some sour fruit(s). However, if the fruits itself is adequately sour, addition of citric acid or any sour fruit is not necessary.

When the pectin extract is added with dehydrating agent to set gel, the precipitated pectin forms insoluble fibre. This is mesh like in structure and can hold (retain) moisture as is done by a sponge. Accordingly, after setting of gel, even if water content is high in the mass, congelation of pectin takes place and it becomes gelatinous (= jelly like), *i.e.*, does not became hard in texture.

In regard to the actual mechanism involved in the setting of gel (= jelly), much remains to be known. A number of hypotheses have however, been put forward to explain the mechanism. Fibril theory is one which suggests that addition of dehydrating substance in the colloidal solution of pectin destabilizes pectin-water equilibrium for which the pectin conglomerates and forms a network of fibrils. The fibril network tends to hold the sugar in their gaps and for this, gel setting takes place. Other opinions have also been put forward and mention may also be made to the Olsen's theory and Hilton's theory. It may be stated that the different postulations which have been put forward are more hypothetical than theories.

4. Jelly-grade

It should be kept in mind that any type of pectin does not in fact, same gel-setting, *i.e.*, jelly-forming capacity. To set gel, some characteristic features of the pectin have necessity and the gel forming capacity of pectin is expressed by jelly-grade of the given pectin sample. By jelly-grade is meant, the amount of sugar that would become necessary for each gram of pectin to form jelly under given environmental condition. That is, it should have high methoxyl content.

Health Benefits of Pectin

Pectin is not only an ingredient to make jelly, jam, marmalade, fruit cheese, pudding, ice-cream *etc.* but this complex polysaccharide has a great role in human health as well. It is recognized as an excellent dietary fibre and thus, its intake into the body plays a vital role in lowering the blood cholesterol level. Apart from binding cholesterol, pectin is also of much help to reduce the triglyceride level in blood, obesity and also redress gastro-intestinal ulcers.

Benefit of pectin in colon, breast and prostate cancer has also been well-substantiated. Necessity of the polysaccharide is felt in pharmaceutical and many other industies.

Chapter 3

Difference between Jelly, Jam and Marmalade

It should be remembered that, the three fruit products, *viz.*, jelly, jam and marmalade are all, basically *jelly.* However, the major differences that exist between them is stated in the following:

1. Jelly

To make jelly, the fruit samples are at first made into fine pieces or the fruit pulp is ground and then boiled with water. If the fruits used are less sour, some amount of harmless organic acid and preferably, citric acid is added during boiling. Instead of acid, some sour fruits such as citrus may also be added.

During boiling, pectin which is present in the fruit samples is extracted in water and addition of acid or sour fruits make thorough extraction of the pectin. On boiling for some time, the pectin comes to the water in the form of colloidal solution. However, along with pectin, other substances which are soluble in water, such as different pectic fractions, vitamins B and C, mineral salts, amino acids, water soluble pigments *etc.* will also come to the extract.

After boiling, the extract is cooled to room temperature. Then it is strained such that the strained extract is clear and does not contain any particle of fruit.

To ensure that fruit particles are not present, straining should preferably be done at first with coarse cloth and then through fine cloth.

The strained out extract is then boiled. During boiling, required amount of sugar is added to the extract. When boiling is completed, it is cooled to room temperature and the semi-solid mass of jelly will be formed due to addition of sugar. Thus, jelly is a gelatinous substance made from fruit extract but this does not contain any fruit particle. If the jelly is taken in a glass container and seen though light, it will be seen that light will pass through it, *i.e.*, it will look translucent and not opaque at any part of it.

2. Jam

Jam is same as jelly but the difference is, fruit pieces remain present in it. In making jelly, straining must be done after boiling the fruit pieces in water. This is done with the idea that fruit pieces are removed out and the pectin extract becomes clear.

On contrary, in the making of jam, straining of the extract is never done after boiling. Hence, fruit pieces remain present in the pectin extract. Thus, jam is a fruit product which unlike jelly contains fruit pieces. Obviously, if the jam is taken in a glass container and seen through light, light will not pass through it and thus, it will look opaque and not translucent as it happens in case of jelly.

3. Marmalade

Marmalade is also a jelly but this product must contain parts of citrus fruits or citrus related fruit such as kumquat. In fact, *shredded peels* of these fruits are mixed in it. To make so, peels of the citrus fruits are at first scraped out with knife. Then the white part of the peel which lies in the inner side and is called albedo, which is botanically mesocarp is removed out by scraping with knife. The outside part of the peel which is green, yellowish or orange in colour depending on species is called flavedo, which is botanically epicarp is taken out for use. This is softened by putting in hot water for some time. After that, the softened flavedo is cut into fine pieces.

The citrus fruits (or the kumquat fruits) are boiled in water after making pieces. When boiling is almost complete, shreds of the peel are mixed in it. The product may or may not be strained. If it is strained after boiling and after straining, the shredded peels are mixed in it and heated for mixing, it will be regarded as jelly-marmalade. On the other hand, if straining is not done and the product is made ready after using shredded peels, the product will be called jam-marmalade.

There may be a number of modifications in the making of marmalade which have been stated under the chapter 7. But it is a universal custom that in the marmalade of any type, shredded peels of citrus fruits must be present. As the marmalade contains shredded peels of citrus fruits and as the peels contain oil glands which have flavour, marmalade is thus, characterized by emission of citrus flavour. However, the product has some bitterness in taste owing to the presence of the peel *i.e.,* oil glands.

The methods of making these products have been elaborated under the Chapters 5, 6 and 7.

Chapter 4

Precautions and Sanitation in Preparation

Before going to make any type of fruit product, the worker should remember the fact that it is intended to be made for consumption of human beings. Hence, it should not contain any such substance that may render harmful effect in the body on consumption. The possible source of contamination of the product and the precautionary measures are stated in the following:

1. Source of Contamination

(a) Unhealthy Fruits

The following biotic and chemical factors could be regarded to be of significance in jeopardising health of fruits.

Biotic Factors

A vast majority of fruit plants belonging to many species are infected by pests that comprise insects, mites, nematodes, fungi, bacteria, viruses or by other animals. Attack by the pests may be on any part of the plant and also in fruits. Damage of fruits by the insect pest, fruit fly is an example which is observed in guava, mango, jamun, ber, pomegranate *etc.* The insect lays eggs on the surface of ripe fruits. On hatching, the maggots go inside the fruits and cause damage. Other fruit borers cause damage of litchi, longan and other fruits.

Symptoms of disease infection on sour lime fruits due to bacterial canker, mildew on grape berries, blue mould symptoms on mandarin orange are commonly noticed. In many bael fruits, symptoms are not visible on rind but on breaking, the pulp is found to be badly infected by black coloured mould. Apart from biological agents, damage of fruits due to physiological disorder occurs in many fruits and black-tip disorder of mango is an example. In the market, many fruits are found to be scratched or half-eaten by birds, bats or squirrels. A fruit infected by pests or diseases or damaged by animal is evidently considered unhealthy and should not be used to any product.

Chemical Factors

Chemical pollution of fruits has assumed a grave situation at the present time in India. This occurs in fruits from four agents, which are fertilizers, pesticides, plant growth regulators and chemicals used for consumers' acceptance in the market.

(i) Indiscriminate use of fertilizers to a fruit plant gives rise to produce which are highly noxious to the body on consumption. It may be stated that methemoglobinemia disease may be a result on consumption of an agricultural produce which received very high level of nitrogenous fertilizer during cropping period. High application of potash to a crop plant is also liable to bring nervous disorder on consumption of the produce.

(ii) Among pesticides, Endosulfan-treated fruits may cause nervous disorder and impaired reproductive functioning, Proxovur-treated fruits may cause cardiac ailment, Chlorpyrifos-treated fruits may bring about thyroid abnormality, Carbaryl-treated fruits may result in nervous disorder and so on. Similarly, too much treatment of a crop with the fungicide, Ziram may result in Parkinson's disease on consumption of the produce and Ferbam may be responsible for renal impairment.

Plant growth regulants, *e.g.*, auxins, gibberellins, cytokinins *etc.* are much used by some fruit growers to control premature drop, improve quality of fruits or increase productivity. Consumption of fruits treated with such chemicals may have adverse effect on health and may be responsible for carcinogenic growth.

Harmful effects may be greater if fruits treated with chlorinated phenoxy acids are used to crop plants.

Some chemicals are applied on fruits after harvest to enhance consumers' acceptance in the market. The hydrocarbon gas, acetylene has much been on use to quicken ripening of many fruits and degreening of citruses, pineapple, peach *etc.*, although use of the gas is not legally permitted for its serious effect on health.

In order to impart desired colour on fruit surface, many dyeing chemicals are sometimes applied before marketing them. It should be noted that intake of fruits treated with such chemicals exert serious hazards in the body. For example, Red-40 treated fruit may cause lymphoma and thyroid cancer Sunset yellow treated fruits may cause adrenal tumour, Erylthrosine treated fruits may impair thyroid functioning, Fast green treatment may act upon bladder and the like. Hence fruits which are treated with such chemicals could not be regarded to be healthy materials.

(b) Unhygienic Materials

The utensils and other materials that are used in making fruit products such as pans, knives, ladle, extractor *etc.*, ingredients and water may also contaminate the products, if they are not used after disinfection. Before use of any equipment, it is therefore, necessary to disinfect them by putting in boiling water for at least ten minutes. Sugar should not contain much impurity and water to be used should be sterile.

(c) Unhygienic Chamber

It is desirable that the processing part should consist of two rooms such that in one room, the fruits, sugar, water *etc.* could be stored while the other room should have necessary installations for cooking and preparation of the products. No unhygienic material should be stored in any of the room and besides that, refrigeration facility in the former room would be of great benefit.

(d) Uncleanliness of the Workers

It should not be forgotten that a worker in dirty clothes and aprons would be a potential source in contaminating the products with harmful microbes and chemicals and thus, making them unfit for consumption.

2. Precautionary Measures

In order to get rid of possible contamination and to prepare the products of hygienically acceptable standard, precautionary measures should be taken at every step of preparation. Heedful attention should in fact, be given at all stages, among which the following are of pertinence:

(a) Procurement of Fruits

Fruit samples for making jelly, jam or marmalade are procured from the markets, fruit growers or other sources. Whatever may be the source, the purchaser should scrupulously sort out the materials and select those on which symptoms of attack by pests or diseases are not apparently visible and are not damaged.

(b) Organically Grown Fruits

Organic farming is a practice in which no synthetic chemical is applied to a crop in its cultivation and the agro-inputs that are applied are exclusively organic materials. Thus, this is a "farming in spirits of organic relationship." Fruit samples obtained from organically run plantations are no doubt, free from any contamination of harmful chemicals. However, organically grown fruits are not obtainable everywhere and where they are available, they are much costly.

(c) Sorting and Grading

On bringing the fruit samples at the processing centre, they are needed to be sorted out at the final stage. This should preferably be done by a panel of persons. Any fruit showing undesirable symptom should always be discarded. Some people have the tendency to make use of a pest infected fruit, cutting off the part which shows symptoms of damage. This could not be recommended as the part of it which looks healthy may also have hidden infection. Hence, it is never justifiable to use a fruit which shows any symptom of biotic infection or chemical spoilage.

Apart from discarding the unhealthy fruits, other sorting are also necessary to separate the fruits according to size and also genuineness of varieties. Many of the small-sized or undershaped fruits in some species are found to develop from unhealthy fruits and these should preferably not to be used. However, it may be mentioned that the "shot berries" in a grape bunch are usually free from infection although their size is very small. Cull fruits borne by many trees are also smaller in size but are not unhealthy in most cases.

(d) Removal of Unwanted Parts

The fruit-stalk, floral remnants at the stylar end, persistent calyx and epicalyx, if any are present in a fruit should be completely removed out leaving no trace. Insect eggs or disease organisms, if detectable in these parts, the fruit should not be used.

(e) Washing

The fruits which have been carefully sorted out to make products are then needed to be thoroughly washed in water to wipe out the dirt and dust that adhere on their surface. Washing should be done by dipping the fruits in clean water and rubbing those with flannel or soft cloth while they are in dipped condition. The washing should be done repeatedly replacing with fresh and clean water and this should be done 4-5 times.

It is apparent that dirt, dust, dyeing chemical, insect remnants, spores of micro-organisms *etc.* are cleared off from fruit surface by thorough washing with rubbing. However, it may be mentioned in this connection that a report of the American Chemical Society (2008) has stated that even after through washing, some bacteria like *Escherichia coli* and some *Salmonella* species remain clinged on fruit surface. But even if these bacteria remain present, they would not cause any harm. (*E. coli* causes infection of urinary tract and gum. Out of 1400 species of *Salmonela*, only *S. typhi* and *S. paratyphi* cause typhoid and para-typhoid fevers respectively but rest of the species have beneficial effect in digesting food in the intestine). As the jelly, jam and marmalade are made on boiling in water, micro-organisims do not survive.

It may be suggested that washing fruits in running clean water is much better than washing in stagnant water. According to B.C. Mazumdar and K. Majumdar (2003), a simple method of washing in running tap water consists of taking a shallow pan, top of which is open and the sheet at the bottom has many perforations. The pan is provided with two convenient handles at opposite sides.

As shown in Figure 3, the fruits are taken in the pan. Holding the handles of the pan, it is brought just below the running tap-water. Tap is turned on and water is allowed to discharge with force. The pan is held and shaken and all the fruits would be thoroughly washed. Strainers and washing devices of other types have also been shown in the Figure 3.

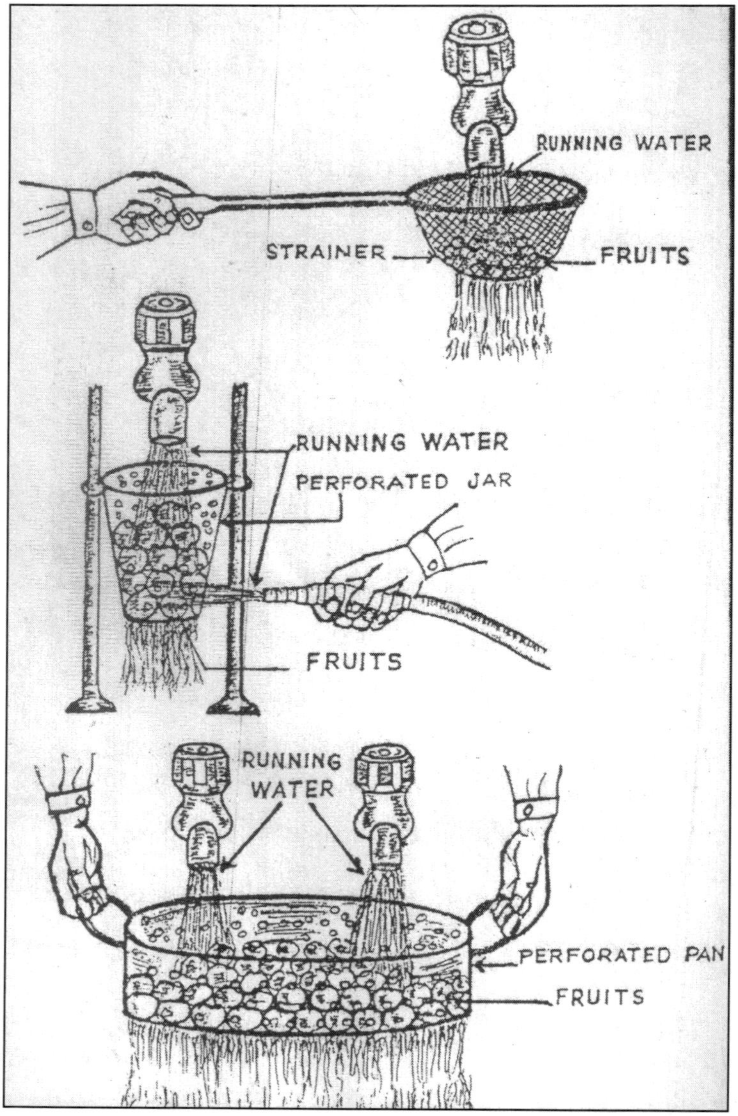

Figure 3. Washing Fruits in Running Tap-water.
(*Courtesey*: Mazumdar, B.C. and Manjumder, K. (2003). Methods on Physico-Chemical Analysis of Fruits. Daya Publishing House).

(f) Making Pieces of Fruits

Knives, secateurs or other implements which would be used for cutting the fruits into pieces should be clean and should preferably be made up of stainless steel. Before use, these should be boiled in water for at least ten minutes.

(g) Disinfection of Storage and Cooking Rooms

The rooms where the products are to be made should at first be made free from dirt, dust, soot *etc.* and preferably, with a suction cleaner. The floor all the walls and the ceiling in each room should also be cleaned. It is advised by many sanitary personnels to spray the rooms with pesticides and close the doors and windows. After a couple of days, the doors and windows should be opened and the fans and exhaust fans should be turned on so that the pesticidal fumes are thoroughly driven out from the rooms.

After waiting for 7-10 days, the rooms should be sprayed with disinfecting chemical and the doors and windows should be kept closed for a day or two. The doors and windows should be opened after that and the fans switched on to drive out the fumes. Waiting for 10-15 days, the rooms should then be used.

To ward off polluting gases, suitable apparatuses are available which makes a room free from smoke and fume within half an hour. Before using the rooms, their walls and floors should be thoroughly washed with clean water and dried.

(h) Washing of Utensils

All utensils and equipments that are to be used should be sanitized and made up of desirable material. Metallic equipments should preferably be made of stainless steel. All equipments are needed to be sterilized before use by boiling in water. Wooden ladles should also be sterilized.

(i) Cleanliness of the workers

The workers should put on thoroughly washed cloths and aprons. They should also cover the hairs with headgears. Coughing and sneezing should not be done in the rooms without use of towels. It should also be remembered that any person clad in dirty clothes should not enter into the rooms.

Chapter 5

Method of Preparation of Jelly

It should be remembered that any straight forward method in the preparation of jelly, jam and marmalade from fruits is perhaps difficult to provide. The principles remaining the same, the actual method(s) involved in making these fruit products varies according to many a factor. The pertinent ones could be stated as species of fruits, their horticultural varieties, maturity condition, infrastructural facility, purity of sugar, preference of consumers and skill of the workers.

However, the general procedure in making jelly from fruits has been outlined in the following steps.

1. Choice of Fruits

It has been stated in the chapter 2 that pectin is the basic ingredient in making jelly, jam or marmalade. When this complex polysaccharide compound in solution (= colloidal) form comes in contact with a dehydrating substance, *e.g.*, sugar, gel setting takes place. It is also necessary that the pectin should be of high jelly-grade, in other words, of high methyl esters. Hence, fruits which are relatively rich sources of high jelly- grade pectin are most suitable materials in making jelly.

Besides pectin, some acidity is also stated by the experts to be a necessary part for the setting of jelly. Hence, those fruits in which pectin is high and have also sourness are in fact, ideally suitable to make jelly.

The following list has given the different levels of pectin and acidity in some fruits.

(i) Both pectin and acidity are high : Oriental pear, quince, crab apple, sour apple, semi-ripe guava, cattley guava, guinea guava, Costa rican guava, feijoa, gooseberry, cape-gooseberry, currant, Japanese plum, cranberry, mulberry, karonda, karna orange, rough lemon, ber, carambola, lemonime *etc.*

(ii) High pectin and low acidity : unripe papaya, fig, mangosteen, rose apple, Surinam cherry, peach, plantain *etc.*

(iii) Medium pectin and high acidity : Sour cherry, cashew apple, blueberry, bael, apricot, monkey jackfruit, semi-ripe mango, diospyros *etc.*

(iv) Low pectin and low acidity : Loquat, raspberry, peach, strawberry, sweet cherry *etc.*

In the choice of fruits for making jelly, the following aspects should also be considered :

Pectin content changes according to maturity of fruits. In early stage of fruit development in most species, protopectin is high and in fully ripe condition, pectin is enzymatically converted into pectic acid which is not suitable in making jelly. In fruits of some species, pectin content and acidity differ even according to varieties, However, fruits which have low pectin content may also be used in making jelly. But in such fruits, other fruits which are rich sources of pectin should be mixed or artificial pectin may be supplemented. Similarly, for fruits of low acidity, other fruits which are more sour may be blended together and used.

2. Removal of Fruit Peel

Peeling of fruits is not necessary for those fruits whose peels are taken along with the fruits. Some examples of such fruits are guava, relatives of guava, apple, karonda, cape-gooseberry, cranberry, ber, carármbola *etc.* Howerver, peeling is necessary to be done if jelly is intended to be made from fruits like pineapple, unripe papaya, unripe or ripe mango, jackfruit *etc.*, peels of which are not edible to the human being.

Peels of citrus fruits belonging to all species consist of two tightly pressed layers. The outer layer is leathery and green, yellowish or orange (may be tangerine) in colour and this is called flavedo. This layer contains plentiful oil-glands which are filled with oil. These oils are highly scented and volatile. The flavor of citrus fruits (also leaves) is due to these volatile oils. (On taking out the peel, if it is bent inward and pressed, oils are found to be thrown out). Under this outer, *i.e.*, flavedo layer, another layer of peel is tightly attached to the flavedo. This inner layer of the peel is botanically mesocarp of the fruit. This layer is soft, cushion-like white or in some cases slightly pinkish in colour and is called albedo.

In citrus fruits of some species, *e.g.*, shaddock, grapefruit, lemon, lemonime and wild citruses such as karna, rough lemon, galgal, amilbed, atoni *etc.*, the albedo layer is thick or very thick and is very soft. It is of interest to note that the albedo layer is rich or fairly rich source of pectin of high jelly-grade. Therefore, excellent jelly could be made from this thick albedo of citrus fruit. But this layer contains plentiful bitter principles, *i.e.*, chemical compounds which are highly bitter in taste. In containing these bitter chemicals, if jelly is made from the albedo layer of citrus fruits, it becomes bitter or extremely bitter and hence, cannot be taken. Nevertheless, it may be mentioned in this connection that many people in the U.S.A. and European countries prefer jelly of bitter taste and such people are fond of such bitter jelly of citrus albedo.

The present author and his associates made experiments to examine whether jelly could be made from the peels of fruits and vegetables of some species. It was observed that peels of mango, pineapple and many fruits gave rise to high quality jelly. Jelly made from mango peel had been of high strength and even difference in the quality of jelly from the peel had been noted according to varieties of mango. Excellent jelly had also been possible to prepare from the peels of onion.

3. Washing of Fruits

This is a vital aspect and this step should be given serious attention. The method of washing fruits has been elaborated under the Chapter 4 (2-e).

4. Making Pieces of Fruits

After washing and surface drying, the fruits should be cut into fine pieces. Pectin is present in the cell-wall of fruits and more a fruit is cut into pieces, recovery of pectin will be higher. Fruits may even be ground in a mixer-grinder and used. Removal of seeds from the fruits is not necessary for such fruits whose seeds are taken along with pulp, *e.g.*, guava, cape-gooseberry, strawberry (achenes) *etc.* but it is necessary for peach, cherry *etc.* In case of citrus fruits, seeds must be removed. This is because, seeds contain chemical compounds which are highly bitter and if they remain present in the jelly, the product will become highly bitter in taste.

5. Extraction of Pectin from Fruit Pieces

Extraction of pectin from the fruit pieces or from the ground pulp of fruits if done in boiling water. Some amount of pectin is extracted in water without heating also but the amount will be less if it is not extracted in boiling water. Besides this, much of the protopectin which is present in the fruits and from which, jelly is not set is in fact, converted into pectin when extraction is done under boiling condition. Hence, for higher recovery of pectin, extraction should be done in boiling water.

For the purpose of boiling, the fruit pieces or the ground pulp of fruits should be taken in a stainless steel or aluminium pan and water should be added to it. How much water is to be added however, depends on the type of fruits. That is, fruits which have high water content, in other words, such fruits which are highly juicy, it should be less and in moderately juicy fruits, it should not be so less. In general, for fruits like guava, pear *etc.*, 700-800 ml of water should be used for each kilogram of fruits.

Boiling in water should be done by mixing with little amount of non-harmful acid and preferably, citric acid. Instead of that however, some fruits and preferably, sour lime or lemon or their juices may be used. Addition of acid is not considered to be an essential feature but according to many pectologists, recovery of pectin from fruits becomes higher in lowered pH condition, *i.e.*, when it is acidified. However, use of acid or sour fruits is not necessary if the fruits are sour.

After addition of acid or sour fruits, the pan should be placed on fire. But before that, its depth should be measured. To do this, a wooden scale should be dipped in it in a vertical manner and the reading should be noted to record the depth of water.

Extraction of pectin in the water in pan is usually completed if heating including boiling is done for 30-40 minutes. It should be noted that too much heating or boiling for long time is not desirable. There is chance of burning of the fruit pieces during heating, *i.e.*, boiling in water. To avoid that, stirring of the fruit pieces with wooden ladle while they are boiling is necessary to be done frequently. When the boiling is completed, the pan containing fruit pieces and water should be removed from the fire and allowed to cool to room temperature condition.

6. Straining of the Pectin Extract

When the pectin extract has been cooled to room temperature condition, it should be strained to remove the fruit pieces and get a clear extract of pectin in water. Straining could be easily done with a coarse cloth or jelly bag (cheese cloth) as shown in the Figure 4.

Figure 4. Straining Pectin Extract through Cloth.

For the purpose of extraction of large amount, it is advisable to use a tripod stand of aluminium or iron. A fairly large thick and strong cloth should be taken and it should be folded to make it more strong. Now, as shown in Figure 5, the folded cloth should be loosely placed on the rim of the tripod stand and firmly tied to that. (A rim of slightly greater diameter may be placed over the rim of the tripod stand such that it tightly holds the cloth in position). The cloth should be moistened with clean water and the pectin extract is slowly poured on the cloth. The colloidal solution of the extracted pectin would be slowly stained out though cloth and this should be collected in a pan. Straining should be done overnight or for few hours to make it completed. The cloth containing the pectin extract should be covered with a cloth piece to avoid coming of insects or dust on it.

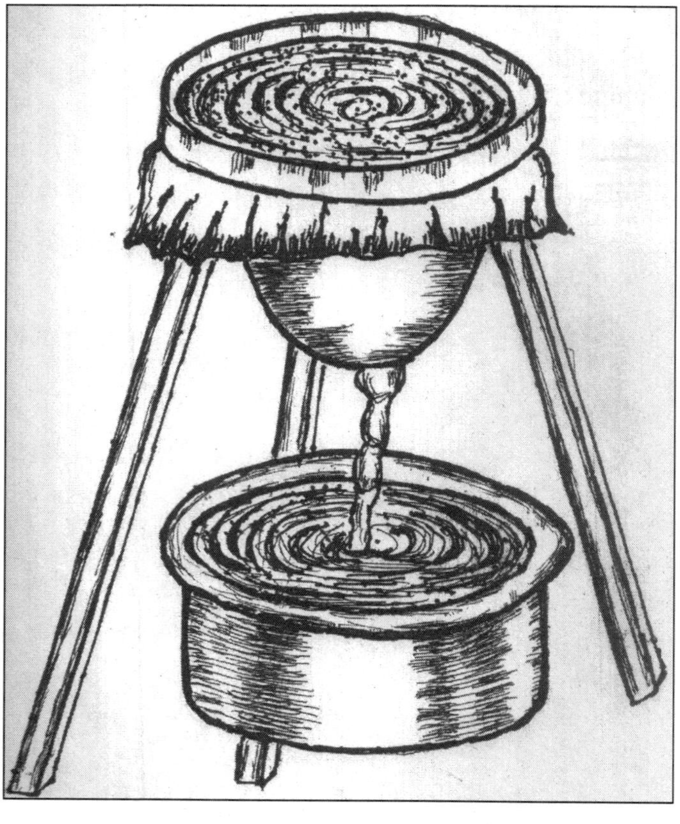

Figure 5. Use of Tripod Stand to Strain Pectin Extract.

7. Determining Requirement of Sugar

In order to set the jelly, the strained out pectin extract must be added with sugar, if it is a jelly for consumption of human beings. The reasons for addition of sugar have been stated in the following:

(i) Sugar is in fact, a dehydrating agent. That is, by dehydrating the pectin in its solution (colloidal) from, it precipitates the pectin. Accordingly, the pectin is coagulated and gives rise to formation of jelly.

(ii) Sugar makes the jelly sweet in taste. It has been earlier stated (Chapter 2: 3) that apart from sugar, there are also many other dehydrating agents but these are poison and are not used in making human consumable jelly. Chemically pure form of sugar, *i.e.*, sucrose is the ideal form of sugar and undoubtedly makes excellent jelly. But this is very costly and is not used to make jelly for household or commercial scale. The ordinary common sugar which is taken by human beings is absolutely suitable to make jelly but it is an important consideration that it does not contain much impurity. It is brought to notice in this connection that many sweetening agents are available in the market, *e.g.*, saccharin, cyclamate, mannitol, isomatt, sorbitol, xylitol aspartame, sucralose *etc.* but these are not dehydrating substances. Hence, although sweet they cannot be used to make jelly as they are not chemically same as sugar.

(iii) By addition of high amount of sugar, jelly, jam and marmalade are protected from the attack of micro-organisms as the osmotic pressure is increased. Hence, high sugar in them acts as a preservative, for which these products could be stored well in ordinary atmosphere and cold storage for them is not necessary.

Sugar is added to the pectin extract (colloidal solution of pectin) in boiling condition in order to form, *i.e.*, set jelly (= gel) when it will be cooled, *i.e.*, comes to room temperature. But how much sugar should actually be added to a given solution of pectin is a very important consideration. This is because, if less than the required amount of sugar is added, jelly will not be set and if more than requirement is added, sugar granules will form in the product and other complications will also arise. Those who are expert in making jelly have adequate knowledge from practical experience in regard to the amount of sugar which should be added to a pectin extract to form an ideal jelly. Many of them are guided by the convention that for making jelly from different fruits, in general, the quantity of sugar which should be used should be same as with the quantity of fruits taken, *i.e.*, for one kg of fruits, one kg of sugar should be used but this would differ to some extent, according to the type of fruits.

It may be brought to light in this connection that there are some methods by which, the amount of sugar to be added to a given solution of pectin could be known in a more precise manner. The two methods which are more commonly employed are stated in the following: These tests are performed taking little amount of the pectin solution while it is boiled after straining.

(a) Spirit Test

To do this test, little amount of hot extract of pectin, *i.e.*, 6-8 ml is taken out with a large spoon or a ladle. This is poured on a dish and allowed to cool. To this, methylated spirit (= commercial grade methyl alcohol) amounting to three times of the pectin solution taken is poured. This is stirred for sometime such that both are thoroughly mixed. The mixture is allowed to settle for some time. Methylated spirit is a strong dehydrating agent and hence, the pectin will congelate and give rise to formation of jelly. This jelly should therefore, be called methylated spirit jelly. (It should be remembered that methylated spirit is deadly poisonous and if a person takes this jelly by mouth, he will die).

The jelly formed from methylated spirit should be carefully examined. If it is found that it has formed in one large lump (Figure 6 a), it should be inferred that the extract contains sufficient amount of pectin of high jelly-grade. For such a pectin extract, 1¼ to 1⅓ part by weight of sugar to the amount of fruit taken should be used. That is, if the amount of fruit taken is one kg, 1¼ to 1⅓ kg of sugar should be added. If it is found that coagulation has taken place in 2-4 lumps instead of one (Figure 6 b), the solution is considered to have medium amount of pectin and for this pectin solution, less sugar should be used, *i.e.*, one kg of sugar for each kg of fruit taken may be added.

On the other extreme, if it is found that coagulation of pectin has not at all taken place by mixing with methylated spirit, or many small lumps have been formed (Figure 6 c), it should be presumed that the solution does not contain enough pectin of suitable jelly grade and hence, the extract should not be considered suitable to make jelly. However, if it is desired to make jelly from this extract, pectin should be supplemented in it. This should be done either by addition of synthetic (artificial) pectin of high jelly-grade, or addition of such fruits which are rich in pectin of high jelly-grade.

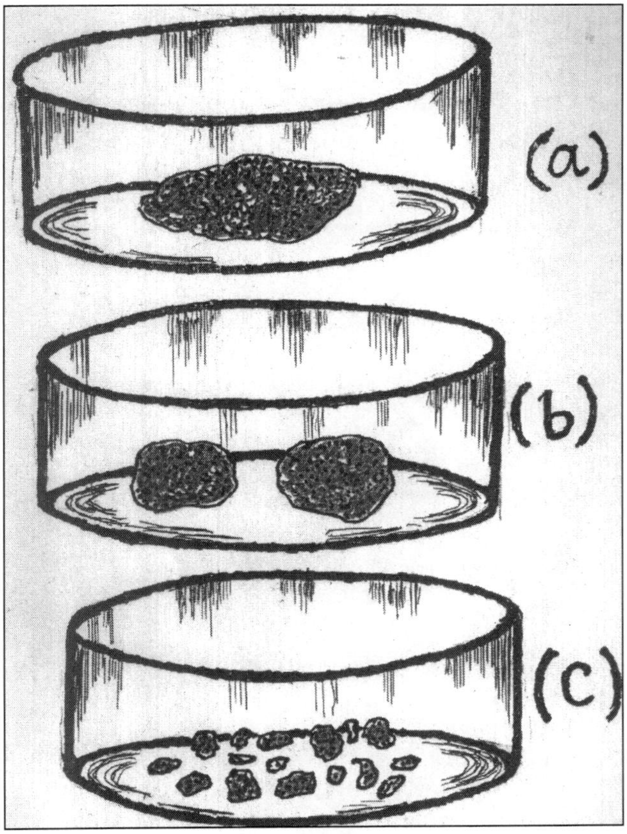

Figure 6. Methylated Spirit Test to Determine Requirement of Sugar. Formation of (a) One big lump, (b) Two small lumps and (c) Many lumps of very small size.

It should be noted that corn starch is also a natural thickener and may be used as substitute of pectin. Gelatin is also considered by many as viable option of jelly. Nevertheless, pectin of high jelly-grade may be considered best and hence, this should be used.

(b) Jelmeter Test

In this test, a small apparatus made of glass is used which is called jelmeter. As shown in Figure 7, this is a graduated glass tube which has openings at both ends. One end of it has broad opening and the opening at the other end is narrow. There are four graduation marks on the jelmeter which are 1¼, 1,¾ and ½.

To use this apparatus, pectin extract (colloidal) should be brought to the temperature of 20° to 40° C. The jelmeter is conveniently held in the left palm in vertical position such that the wider opening is at the upper end. Holding the tube in such position, its narrow opening at the lower end is tightly closed with the small finger of the same palm. As shown in Figure 7, pectin extract should be slowly poured into the jelmeter through its upper (wider) opening upto the brim to fill it up with the help of a spoon.

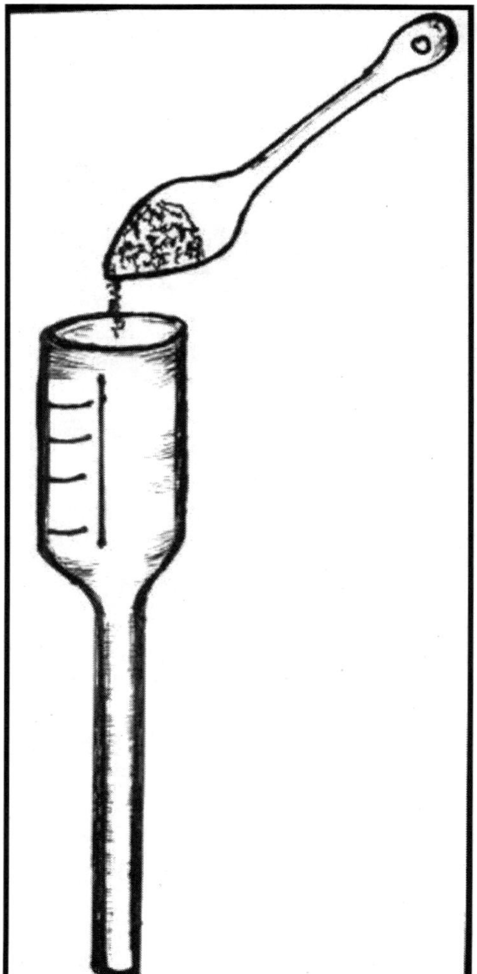

Figure 7. Jelmeter.

Now, the small finger should be removed so that the pectin solution in the jelmeter flows outside through the lower opening of it. Time should be noted with a watch and exactly after one minute, the lower end should be closed with the small finger such that the flow is stopped. On closing the lower end with the small finger, the marking on the Jelmeter where the pectin extract stands should be noted. That is, marking which is nearest to the surface of the pectin extract should be considered. From this reading, the amount of sugar which is to be used to the pectin extract is known. For example, if the pectin extract stands at a position whose nearest upper marking is 1¼, then 1¼ cup of sugar should be added for each cup of the pectin extract.

If it is observed that the entire pectin extract passes out through the lower (narrow) opening and nothing is left in the Jelmeter just after one minute, it should be presumed that the extract does not contain adequate pectin to make jelly from it. In such case, the extract is needed to be enriched with pectin by adding synthetic pectin or from other fruits that are rich sources of pectin.

8. Addition of Sugar to the Pectin Extract

After knowing the amount of sugar which is to be added to the pectin extract, it should be weighed and kept aside. An important point to remember is, sugar should as far as possible be pure and does not contain much impurity. Various types of sugar are actually available in the markets, among which, that is pure, white in colour with very small sized crystals, not adulterated and does not contain dust or dirt should be chosen.

Sugar should be added to the pectin extract while it is in boiling condition and when sufficient quantity, *i.e.*, about ¾th of the solution has evaporated. To do so, the depth of the pectin extract in the pan should be known before the pan is placed on fire. It should is done by vertically dipping the same wooden scale which was dipped to measure the depth before the pan was placed on fire and stated under the Heading 5. By dipping that scale at suitable intervals into the boiling extract into the pan, the depth should be known. On doing so in this way, the amount of water evaporated from the colloidal extract of pectin would be known. Thus, when it will be seen that ¾th of water has evaporated, *i.e.*, the reading in the scale has come down to ¼th level, sugar should be added to the boiling extract of pectin.

It should be noted that the sugar should be added slowly and not the entire amount at one time. It is also important that the boiling extract should be constantly stirred with ladle while addition of sugar or else, the sugar will burn and there will be sugar burning smell in the jelly. After addition of sugar, it will be seen that scum will appear at the surface of the solution. This is due to the impurity of the sugar and should be regularly removed out with ladle. (Whatever pure sugar is used, the scum will appear if it is common sugar but will not appear if chemically pure sucrose is added, which is however, difficult to use because of very high cost).

9. Determining Finishing of Heating (= End-point)

Too much heating of the pectin extract is not necessary after addition of sugar to it and the pan should be removed from fire. If too much heating is done, the jelly will be very hard and sugar crystals will form in the jelly to make it a granular jelly. Hence, the proper time when the pan should be removed from fire should be known. This is called end-point, or finishing point and this may be known easily from any of the following two methods:

(a) Spoon Sheet Test

To perform this test, little quantity of the pectin extract which is in boiling condition should be taken in a metal spoon. After that, the spoon containing the extract should be partially cooled but not fully cooled to come to room temperature. While the extract is slightly hot, the spoon should be turned at one side as shown in Figure 8. The extract will be seen to run off at the side of the slanting spoon.

If it is found that the pectin extract tends to flow down from the spoon in a hanging manner, *i.e.*, as flake (sheet) as shown in Figure 8 a, it should be presumed that the end-point has reached for jelly and the pan containing the pectin extract should not be heated anymore and it is to be removed from fire. But if it is found that instead of hanging like flake, the extract runs off in broken flakes (Figure 8 b), it should be considered that the jellying point has not come and the extract should be heated for some time more. Test should be done later and heating continued until the extract is found to hang from the spoon as in case of Figure 8 a.

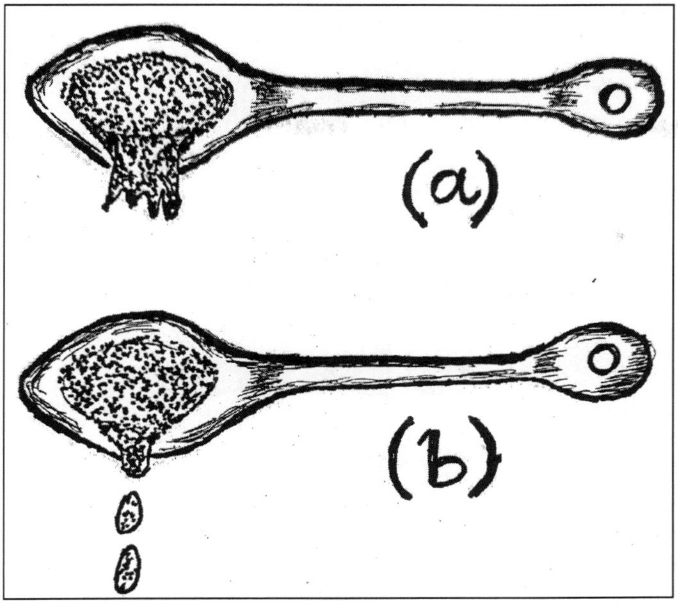

Figure 8. Spoon Sheet Test to Determine Jellying Point.

(b) Plate Test

To do this test, little amount of boiling extract of pectin extract is taken with a spoon and poured on a small dish (Figure 9). It should be cooled to room temperature condition. After some time, if it is found to have well coagulated and is broken on pressing with a finger, it should be considered that the pectin extract has attained the required jellying point and in such case, no more heating of the extract should be done. On the other hand, if it is found that the lump has not adequately coagulated, trickles down and is soft such that on pressing a finger it sinks, the jelly should not be considered to have reached end-point and further heating of the extract is necessary. Similar tests should be done after heating for some time until the result comes to the former type.

When the end-point, *i.e.*, jellying point (= gel-setting point) of the pectin extract is reached, it is considered that the jelly has been prepared. In hot condition, it is therefore, transferred to suitable containers.

(10) Disinfection of Containers (Glass jars)

Most commonly, jars made of glass are universally used to store jelly, jam or marmalade, although aluminium cans are sometimes found to be used but rarely. Plastic containers may be used to store jelly but in any case, hot solution of jelly should not be stored in a plastic container. However, transparent glass jars could be said to be best and jelly stored in such jars could be visible from outside whether any abnormality has resulted in it during long term storage.

Figure 9. Plate Test to Determine Jellying Point.

Before pouring the hot, liquid jelly into the glass jars after its removal from fire, it is a must that the jars should be disinfected. This is because, if the jars contain micro-organisms, they would multiply and grow during storage of the jelly in the jars and may destroy the product.

The disinfection of the jars is best done by putting in boiling water for some time. Use of chemicals to disinfect them should be avoided.

To disinfect the glass jars, a large aluminium or stainless steel pan should be taken. Several layers of towel or flannel should be placed at the bottom of the pan to act as false-bottom. The jars as well as their lids are at first thoroughly washed with soap water, then with clear tap water and them they are carefully placed over the false-bottom which is kept in the pan.

Clean water should then be poured into the pan near upto the brim such that all the glass jars and their lids remain dipped in water. The pan containing the jars, their lids and water should then be placed on fire to boil the water

(Figure 10). The water should be allowed to boil for about ten minutes. Owing to such boiling, the materials will be disinfected. (The false-bottom of towel or flannel cloth which has been kept at the bottom of the pan would protect the glass jars from breakage).

The disinfected jars should then be carefully taken out from hot water with suitable tongs and the water in them should be completely thrown away. In order to avoid breakage of the glass jars, it is advisable to place them on a wooden board which should be dry and any drop of water should not be stuck on its surfaces.

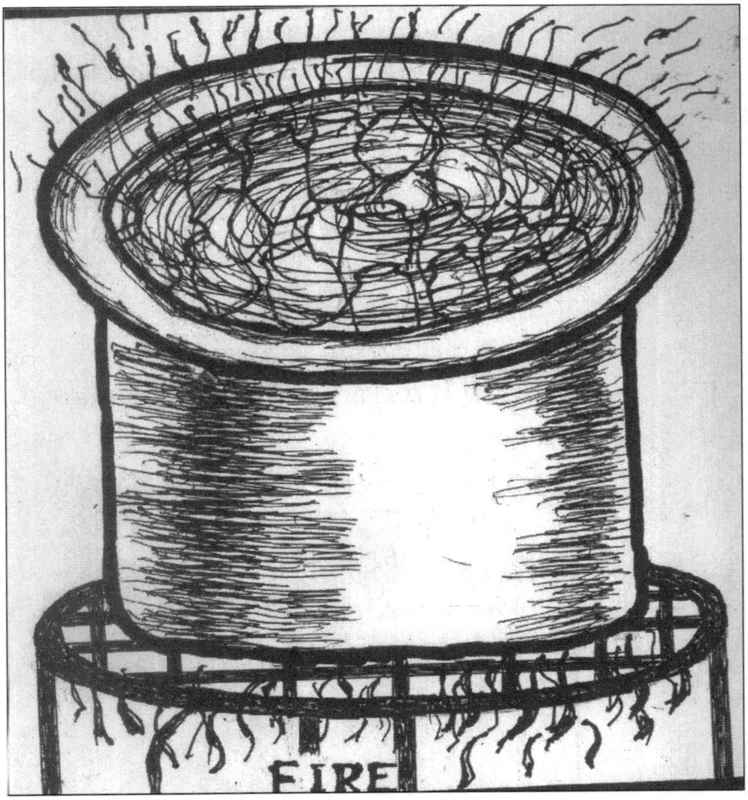

Figure 10. Disinfection of Glass Jars and Lids by Maintaining in Boiling Water.

(11) Filling Glass Jars with Hot, Liquid Jelly

The glass jars kept on wooden board should then be filled with the liquid form of the jelly on removal from fire. This is an important operation and should be done very carefully and slowly. This is because, if the pouring of the hot, liquid jelly into the jars is attempted in a hurry, there is every chance of getting air-bubbles to be trapped inside the jelly, and the jelly on solidification will have a hazy appearance and not translucent. Hence, pouring of the hot, liquid jelly should be done slowly and part by part with suitable time interval to avoid entry of air-bubbles.

(12) Solidification of the Jelly (= Semi-solid Form)

The hot jelly which is poured into the jars is in liquid form. This is needed to be solidified for use. The liquid form of the jelly is naturally solidified under room temperature condition within 12-14 hours. Maintaining in somewhat cold condition gives better result but it should not be done in a refrigerator as the glass jars will crack on doing so. The jars should not be covered with lids when allowed to solidify. However, the glass jars containing the jelly should be covered with clean cloth to avoid entry of insects and dust.

(13) Covering with Wax

If the jelly maintained in the jars is stored for some months, it is often found that a thin layer of water appears on the surface of it. Owing to this, part of sugar also comes to the surface and for this, micro-organisms grow on the surface which later would go inside. To avoid this undesirable condition, the jelly in the jars should be provided with a layer of praraffin wax. To do this, paraffin wax should be melted and very little quantity of the molten wax should be poured on the jelly kept in the glass jars. This should be done when the jelly has thoroughly solidified and has come to room temperature. When the jelly should be taken, the thin paraffin layer should be removed and this could be easily done with finger or spoon.

(14) Labelling of the Product

It is a necessary part that every jar of jelly should be labelled stating at least some particulars such as name of fruit used, date of preparation and the name of the person who prepared it.

(15) Storage

High amount of sugar is added in making jelly for which, its osmotic pressure is greatly increased. Accordingly, micro-organisims do not attack the jelly and it is not necessary to store the jars in refrigerator. However, although microbial infection does not occur but physiological damage of many types is seen in storage and these have been discussed under the Chapter 8.

Chapter 6

Method of Preparation of Jam

It has been stated earlier in the Chapter 3 that jam is basically a fruit jelly but unlike jelly, fruit pulp remains present in it. As it is made by boiling of the pieces of fruits, they are automatically ground to form a pulpy mass. In a jam, original size and shape of the fruits are not retained. But a type of jam is made, which is known as "jam preserve", in which the actual appearance of the fruits are maintained. Such type of jam has much popularity in the east European countries.

Most commonly, jam is made by mixing fruits of two or more species and such types of jam is called "mixed fruit jam." Sometimes it is made by adding kernels of nuts of cashew, walnut, pista, hazel nut, pecan nut *etc.*, or dates, raisins, prunes or other dried fruits. Such types of jam is usually referred to as "conserve" or "jam conserve".

The general procedure involved in making jam is stated here but there may be many modified forms also.

1. Choice of Fruits

The same species of fruits which are used in making jelly and have been referred to in Chapter 5 (1) are suitable in making jam either using them alone or more commonly, by making a combination of two or more types of fruits.

Those fruits which have some *flovour* usually get preference in making jam rather than jelly from them. Some fruit preservation companies attach much importance to some particular fruits in making jam and among these, mention may be made to pear, apricot, raspberry, blackberry, cranberry, pineapple, cape-gooseberry, cashew apple (= swollen peduncle and thalamus of cashew tree), plum, gooseberry, loquat, mango, strawberry, peach *etc.* The common practice is to make the jam by combining fruits of two or more species of fruits but this is not essential.

The procedure for washing and peeling of fruits and making them into pieces to make jam would be same as has been described in the method of preparation of jelly (Chapter 5: 2 and 3)

2. Lessening Sourness of Fruits

It has been stated under the Chapter 2 and 4 that according to many expert pectologists, sourness has its necessity in extraction of the pectin from fruits and for higher recovery of pectin. Accordingly, a non-harmful organic acid and preferably citric acid should be added during boiling but this is not necessary if the fruits are already sour.

The above principle is true in making jam also. Hence, lowering of pH level by adding sour fruits or citric acid has necessity if fruits are not adequately sour. But it may be mentioned in this connection that people and children who are fond of jam have a liking of such type of jam which has negligible sourness. Hence, if much of citric acid or sour fruits are used in extraction of pectin, the ultimate product becomes very sour and does not gain much favour. Hence, acid or sour fruits should be used in moderation.

In this connection, it may be stated that if the fruits which are to be used to make jam are very sour, it is desirable that the acidity should be lessened in them. To do so, clean water should be taken in a container and little amount of lime should be added to the water and vigorously shaken to make an emulsion.

The fruit pieces should be dipped in the lime water and thoroughly shaken. Lime water is alkaline and hence, it would lessen much of acidity present in the fruit pieces. After dipping in lime water for 5-10 minutes, the fruit pieces should be repeatedly washed in clean water, making change of the water or in running tap water so that no lime particle adheres on fruit surface.

3. Pulping of Fruits

Jam is prepared by boiling the fruit pieces in water with addition of required amount of sugar in it. During boiling, the fruit pieces are naturally broken, ground to form pulp and jam is formed by the mechanism of gel setting. However, for such fruits which are hard in texture, *e.g.*, unripe mango, unripe papaya *etc.*, thorough grinding so as to form pulp usually does not take place during boiling.

For fruits of such type, it is advisable to grind the pulp artificially before boiling in water. To do this, the fruits should be slightly warmed with water to make them softened. While they are hot, their peels should be removed and they should be rubbed on a wire net of steel or aluminium placed over a container. Rubbing may be done with the help of a stone chip and the ground pulp which is sieved through the wire net is collected in the container. This should then be boiled in water to make jam.

4. Addition of Water and Boiling

The fruit pieces or the ground from of it obtained by sieving on wire net should be weighed and this should be taken in a stainless steel or aluminium pan. Water should be added to the pan to make jam. The amount of water which should be used should be relatively less than that which is used to make jelly. In general, for each kg of fruits, 600-700 ml of water should be added to make jam. This would however, depend on the type of fruits such that, amount should be less for fruits having high content of juice and more for fruits if juice is less.

On addition of water, the depth of it should be measured by vertically dipping a wooden scale as is done in case of making jelly (Chapter 5: 8) and the reading is noted.

The pan containing the fruit pieces or ground pulp and water is then placed on fire to boil. The fact should be well-remembered that although in making jelly, the extract of fruit pieces in water is strained after boiling to remove out the fruit pieces and get an extract of pectin, it should *never be done* in making jam. That is, fruit pieces will be thoroughly ground due to boiling and it will remain present in water. Heating and boiling should be continued and at suitable intervals, the depth of the boiling liquid should be examined by dipping the same wooden scale as is done in making jelly. When it will be observed that about ¾th of the water has evaporated, sugar should be added to the pan while it is in boiling condition.

5. Addition of Sugar

The amount of sugar which is necessary to be added to make jam should be determined in the same way as stated in the making of jelly (Chapter 5: 7). Nevertheless, unlike jelly, some higher amount of sugar is usually added in jam to impart high sweetness. Regarding addition of sugar, similar procedure as stated in the Chapter 5(8) should be followed. If "conserve" is to be made, kernels of the nuts of cashew, walnut, almond *etc.* or date, raisin or other dry fruits should be added to the pan while the extract is boiling and few minutes before the pan is taken out from fire.

Determination of the finishing-point, *i.e.*, end-point in making jam is more critical than that of jelly. It should be noted that texture of jam is less hard than jelly. The procedure in regard to disinfection of glass-jars, filling them with hot and liquid form of jam, their solidification, providing wax layer on their surfaces labeling of jars and their storage should be same as described in the making of jelly (Chapter 5: 9-15).

Chapter 7

Method of Preparation of Marmalade

The word, marmalade has derived from marmalade, a Galician-Portuguese term. As specified in the Chapter 3(3), marmalade is basically a jelly or jam but it must contain peels obtained from different species, varieties or hybrids of *Citrus* fruits or species which are closely related to those of *Citrus, e.g., Fortunella* or *Poncirus* species. However, in commerce, marmalade from mandarin orange is more common at least under Indian condition. The most important point is, fruits of whatever species is used in making marmalade, pieces made from the peels of citrous (= citrus related, i.e., fruits of the genera, *Fortunella* and *Poncirus*) fruits must be present in it.

It has been stated in the Chapter 5(2) that peels of citrous fruits consists of two layers. The outer layer which is leathery and green, yellowish or orange in colour according to species is called flavedo (= epicarp). The inner layer which is tightly pressed with it and white or slightly pinkish in colour with cushion like structure is called albedo (= mesocarp).

The flavedo contains plentiful *oil-glands* and the flavour of citrus fruits is due to the volatile oils that are present in these oil-glands. These volatile oils give rise to the characteristic flavour of citrus in marmalade. However, the volatile oils are also bitter in taste and due to that, taste of marmalade is a little bitter.

It is observed that marmalade is not so popular to the majority of the Indian people than jelly and jam and this may be due to the fact that it has some bitter taste. But it may be mentioned that the bitterness is rather a likeable feature to the people of the western countries and it is in fact, much favoured to them.

Marmalade may be of two types. These are called, (i) Jelly-marmalade and (ii) Jam-marmalade. In jelly-marmalade, the extract of citrous fruits is strained out to remove the fruit pieces and then peels of the citrous fruits are added to the strained extract. On the other hand, no straining is done in making jam- marmalade and thus, parts of fruits but not their seeds remain present in jam-marmalade. It may be stated that, jelly-marmalade has greater acceptance to the consumers than jam-marmalade. There are various methods to prepare this and a method, which is commonly followed is described below.

1. Choice of Fruits

In making marmalade, citrus or citrus-like, i.e., citrous fruits belonging to two or more species are usually combined together and used*. Preference is given in selection of fruits of such species, which have firstly, high flavour and secondly, albedo layer of the peel be somewhat thick but not too thick. In fact, thickness is given importance as abledo layer of citrous fruits is rich in pectin of superior jelly-grade. For this reason, some part of the albedo layer is kept to make the marmalade. On the other hand, albedo layer has much bitter chemical compounds as well and hence, the marmalade may be highly bitter in taste if too much of them is used. Accordingly, citrus fruits whose albedos are very thick, e.g., shaddock, grapefruits, atoni, etc. are not used making marmalade if not, specially desired to make so.

Citrus fruits which are more commonly used in India to make marmalade are mandarin orange and some lemon varieties. In this connection, it may be mentioned that fruits of rough lemon (= Jambhiri), Karna, hill lemon (= Galgal), the tangor variety, Kinow (= a hybrid of the sweet orange variety, King orange and the mandarin orange variety, Willow leaf, some tangelo varieties *etc.* are highly suitable to make marmalade.

---

* (N.B.: It should be noted that the term, citrous refers to botanical species which come under the genera, *Citrus*, *Fortunella* and *Poncirus*).

High quality marmalade could also be made from the meyer lemon and the lemon varieties, Malta and Nepalese oblong. Marmalade prepared from Ponderosa lemon (= hybrid of *Citrus maxima* and *Citrus medica*) has high popularity in western countries. The famous lemon variety of West Bengal, known as Gandharaj has highly pleasing flavour and is much used in West Bengal to make marmalade in combination with fruits of rough lemon variety, Jambhiri.

Fruits of the *Citrus* species, known as Ada jamir which is grown in Assam and is characterized by distinct smell of ginger in the fruits is also merited for making marmalade in eastern India.

Excellent marmalade is made from the fruits of a wild *Citrus* species which is know as Calamondin. Fruits look like mandarin orange in colour, semi-loose skinned, very small in size and of highly sour juice. Trees of this *Citrus* species are not very tall and grown in ornamental gardens and even in large pots. The tree goes by the name, Hazara in many parts of India for it is stated that a tree produces one thousand fruits in season, although this is not true. Apart from fruits belonging to the *Citrus*, high quality marmalade is also made from kumquat fruits which are closely related to citrus fruits.

On procurement of *fruits*, they should be washed as stated under the Chapter 4(2-e). But the citrous fruits which should be used to make marmalade should not be cut into pieces as is done in making jelly or jam.

2. Preparation of Peels

This is an important step in making marmalade. It should be remembered that marmalade gains favour for the characteristic flavour of citrus or citrous fruits and that the outer layer of peel, *i.e.*, flavedo contains oil-glands. The oil glands contain volatile oils which are actually responsible for flavour of citrus or related fruits. Hence, the flavedo is taken out, *i.e.*, separated out from the albedo. It should be carefully done with a knife and may also be done by scraping to make sure that this does not contain even very small part of albedo. (If the flavedo which has been separated out is seen against light, plentiful oil-glands, small and large could be distinctly seen in it). On the other hand, the albedo does not contain oil-glands but it contains many chemical compounds which are *bitter in taste.* Accordingly, the marmalade would be highly bitter in taste, if much of albedo comes in it.

On separating out the flavedo, it should be made into small pieces and to make so, it is needed to be softened. The softening of the flavedo should be done by warming it for some time, placing in water. The softened flavedo should then be made into pieces, *i.e.*, shreds. It should be done with a suitable knife and in such a manner that each shred becomes 2-3 cm long and 2-5 mm wide. To impart highly pleasant flavor, sometimes peels (= flavedos) of highly scented fruits of different *Citrus* species are also used. (= Some manufacturing companies add lemon oil to the marmalade to bring about intense flavour of lemon.

If necessary, the shredded peels of flavedo could be stored for prolonged period such that they might be used later. For such purpose, special preparation of the flavedo becomes necessary as stated below. The peels should be dipped in saline water and the concentration of the saline water should be gradually made increased. That is, at first they should be dipped in 2 per cent water solution of common salt for 1-2 hours. Then they are taken out and dipped in 4 per cent salt solution and kept for an hour. After that, they should be transferred to 6 per cent and in this way, upto 15-20 per cent salt solution. Little quantity of vinegar should be added when stored in 20 per cent salt solution.

Then the flavedos should be taken out and shredded into pieces. It should be done in such a manner that each shred becomes 2-3 cm long and 2-5 mm wide. These shredded peels should be placed in warm water for some time to remove the salt. The salt-free shreds are then stored in jars in 50 per cent solution of common sugar and little amount of citric acid should be added to it. The shredded peels could be stored for some period or months by this processing but some flavour might be lost in them during storage.

3. Extraction of Pectin by Boiling in Water

The citrus or related fruits which are to be used to make marmalade should at first be made free from flavedos by cutting off them. The albedos should be left as such, if they are not very thick. But if the albedos are very much thick, it becomes necessary to cut off some parts of them. It should be remembered that albedo of citrus or related fruits is rich in pectin of high jelly-grade and hence, it has much importance in making marmalade (= formation of gel). But at the same time, albedo contains chemical compounds which are bitter in taste. Accordingly, if much of albedo is retained in a citrus fruit in making marmalade, the marmalade will be highly bitter in taste. In making marmalade, some parts of the albedo, should therefore, be retained but not much.

After removing the flavedo completely and albedo partly, the fruits are to be cut so that a fruit is made into parts. Then the seeds present in each fruit should be completely removed with spoon or a coring knife to make sure that even a single seed is not present in any fruit. It should be remembered that seeds of all citrus fruits are extremely bitter and presence of even one seed would make a marmalade highly bitter in taste.

On removal of seeds, the cut out fruits should be boiled in water, taking them in aluminium or stainless steel pan. Amount of water which is to be added to the pan for boiling (= pectin extraction) varies according to fruit but in general, not more than 500 ml of water becomes necessary for a kilogram of fruits. If jelly-marmalade is to be made, 500 ml of water is to be added for each kg of fruits and it should be 400 ml per kg of fruits if jam-marmalade is to be made.

It is recommended that addition of acid becomes necessary for thorough extraction of pectin. But as the citrus fruits which are used in making marmalade are already sour, any further addition of acid does not appear to be necessary. Nevertheless, in the preparation of marmalade from mandarin orange and sweet orange, little amount of citric acid is most commonly added as these fruits are not much sour in taste.

After addition of water and acid (or sour citrus fruits) in the pan, it should be placed on fire. But before placing the pan on fire, it is an important step that the depth of water in the pan should be measured. This should be done by dipping vertically a clean wooden scale upto the bottom in the way as has been stated in making jelly (Chapter 5: 5).

If jelly-marmalade is to be made, the extract should be strained as is done in making jelly and described in the chapter 5 (4). But straining should not be done and the pulp will remain in the product as is done in the making of jam (Chapter 5: 7).

4. Determining Requirement of Sugar

This should be determined following the method as has been described in the making of jelly under the Chapter 5 (7). However, it should be noted that somewhat higher amount of sugar is added in making jam and marmalade than in the making of jelly.

5. Addition of Sugar to the Extract

The same principles should be applied in adding sugar as has been put forward in the Chapter 5(8). Sugar should be added to the extract while it is in boiling condition and when ¾th of the water has evaporated. To know whether ¾th of water has evaporated, the same wooden scale which was initially dipped before start of boiling as has been stated under the Heading 3 in this Chapter should be dipped into the boiling extract at suitable intervals (Chapter 5: 8). When it would be known that ¾th of the extract has evaporated, sugar should be added. Addition of sugar should be done part by part and the extract should be constantly stirred (Chapter 3: 5) during addition of sugar.

6. Addition of Shredded Peels

The shredded peels are to be added to the boiling marmalade in the pan after about 15 minutes of addition of sugar to it. The amount of peels which should be added varies and some people prefer to add higher amount while addition of lesser amount is preferred by many. In general, 350-400 gram of shredded peels are added to prepare marmalade from one kg of fruits. Addition of relatively high amount of peels is usually done in jelly-marmalade than in jam-marmalade.

Immediately after addition of the shredded peels, the boiling marmalade should be vigorously stirred with a wooden ladle, such that the peels are evenly distributed in the boiling mass. This is because, often it is found that many of the peels have on floating reached the upper part and the surface. This is not desirable and to ensure that they are uniformly distributed in a homogenous manner, vigorous shaking of the boiling material both up and down and sidewise with the wooden ladle should be done for some time. The boiling marmalade containing the shredded peels should be removed from the fire after about 10 minutes of addition of the peels.

Determination of end-point (= finishing point, marmalade – setting point) may be done following the method as stated in making of jelly (Chapter 5: 9). Rest of the procedure should be done as described in the Chapter 5: 10-15.

Chapter 8

Faults of Jelly, Jam and Marmalade and Corrective Measures

After the jelly, jam or marmalade has been prepared, sometimes it is found that these products have not been made upto the desired standard and hence, could be regarded to be faulty. Abnormalities may also appear when these products are stored in jars. A number of such type of defects are commonly met with and the more common ones have been brought down in the following:

1. Improper Setting

Jelly, jam and marmalade are semi-solid, gelatinous (= gluey) mass which are formed (= set) on coagulation, *i.e.*, congelation of pectin. If gel setting has been properly done, the jelly, jam and marmalade which have been stored in the jars would not roll down if the jars are slanted. But in case of improper setting, it may be seen that these would tend to flow down in the jars.

In such case, these products should be taken out and then they should be boiled again by adding some more pectin. To do so, *i.e.*, to supplement pectin, they should be boiled with water and fruits which are rich in pectin of high jelly-grade as has been stated under the chapter 5(1). However, artificial (= synthetic) pectin, *i.e.*, already extracted pectin (as stated in Chapter 9: 1) may be procured from a chemical or druggist's shop and used to supplement pectin.

2. Very Hard in Texture

Texture of jelly, jam and marmalade is a very important consideration and it greatly determines quality of the products. It should be noted that texture should be ideal, if it is neither very hard nor very soft such that they can be taken out with an ordinary metallic spoon giving clean cut.

Very hard textured jelly, jam or marmalade results for some reasons. These are, if the pectin is added in excess amount, if boiling is done for long time, if the fruits used are not adequately mature *etc.* The stiffness of the jelly, jam or marmalade could however, be softened to bring them to normal texture by taking out from the jars and boiling again on addition of some water.

3. Very Soft in Texture

It should be noted that texture of jam and sometimes marmalade is less hard than that of jelly. However, undesirably soft textured jelly, jam or marmalade should be considered to be a fault. On leaning the jars containing the products of such type, it will be found that they tend to roll down easily. If metal spoons are placed on them, those will sink easily into them without being pushed.

Lack of adequate pectin has been the cause of such undesirableness. To correct this, same procedure should be adopted as stated under the Heading 1 in this Chapter. It may however, be mentioned in this connection that soft-textured jelly, jam or marmalade are sometimes preferred by some consumers as they could be easily spread on bread slices.

4. Accumulation of Water

It is often found that when the jars containing jelly, jam and marmalade are stored for some months, water appears on their surface. Mistakenly it appears that water has fallen on their surface. The abnormality is stated as "weeping" jelly, jam or marmalade. The condition appears due to hydrolysis (= chemical breakdown) of pectin in the products. Addition of too much quantity of citric acid or sour fruits during preparation of these products is a common cause for the hydrolysis of pectin. Hence, addition of excess acid during boiling should be avoided to avert this abnormality.

5. Turbid Appearance

If the jelly is properly done and kept in a jar, it should have transluscent appearance. That is, if a jelly-filled glass jar is seen through light, light will be found to pass through it. But sometimes it is found that the jelly is cloudy, *i.e.*, hazy in appearance when seen against light. This is common in jam and marmalade also though not easily detectable.

Most common cause of this abnormality is trapping of air-bubbles. This happens particularly when the boiling jelly, jam or marmalade is poured into the glass-jars hurriedly. If pouring is done swiftly, air-bubbles go inside the products and they are trapped in them and thus, gives rise to cloudy appearance. Hence, pouring into the jars should be done slowly, part by part and in slantwise position of the jars.

Another cause is, if particles of fruits pass through during straining (Chapter 5: 6). To avoid this, straining should be done twice, at first with coarse cloth and then the strained extract should be strained again through fine cloth. Still another cause may be there, that is, if the fruits contain high amount of pectin of high jelly-grade, the jelly, jam or marmalade coagulates (= congeals) quickly, for which air-bubbles are trapped.

Removal of air-bubbles from the glass jars could be done on warming. To do so, folded cloth or flannel should be placed at the bottom of a stainless steel or aluminium pan to act as a false-bottom. Over this, the glass jars containing the products should be vertically placed opening the lids such that, open ends of the jars face upwards. Water should then be poured into the pan so that, half of the jars remain dipped.

The pan containing the jars and water should be placed on fire to act as a water-bath and slowly heated but water should not be allowed to boil. After sometime, the pan should be removed from fire and water in it cooled to some extent.

The jars containing the products should then be placed on a dry wooden board (Chapter 5: 10). When they will be cooled to room temperature, thin layer of paraffin wax should be provided on their surface (Chapter 5: 13).

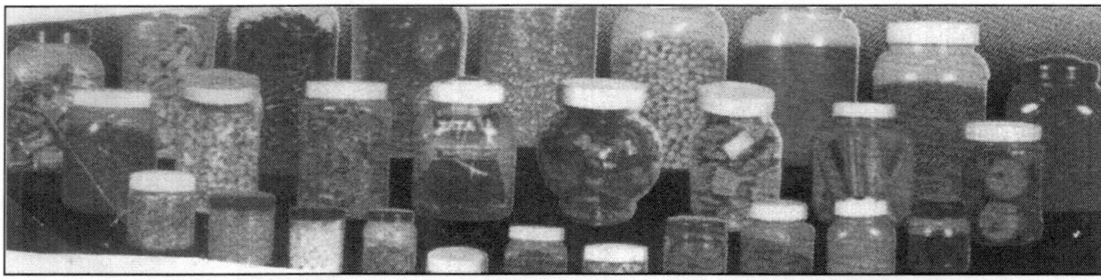

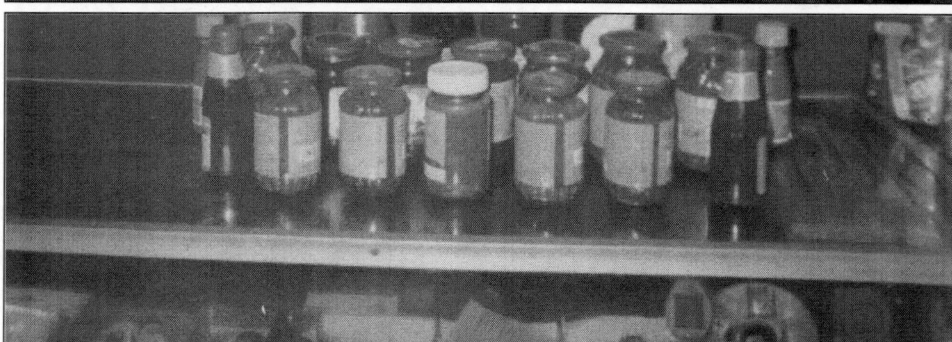

6. Sticky Form

After storing the jelly, jam or marmalade in glass jars for a month or two, it is sometimes found that they have assumed a sticky appearance. On touching with fingers, they give a gluey or adhesive feeling. Such a condition arises when the products are made with inadequate amount of pectin. To correct this undesirableness, pectin should be supplemented in the products following the procedure as stated under the Heading 1 in this Chapter.

7. Presence of Sugar Granules

Presence of sugar granules is sometimes met with in jelly, jam or marmalade. The granules are particularly felt when the products are given in mouth. Inadequate amount of pectin is usually the cause for this. If such condition is felt, the products should be taken out from the jars and prepared a fresh on supplementing on afresh supplementing pectin as stated under the Heading 1 in this Chapter.

8. Brittleness

Brittleness of jelly, jam or marmalade after storing them in glass jars for some period may sometimes be an occurrence. It is observed that when the products are taken out from the jars, they come out in broken fragments. If this type of abnormality is found, the products should be taken out from the jars and should be prepared again by heating with addition of some common salt and citric acid.

9. Fermentation

Fermentation of the jelly, jam and marmalade stored in glass jars is often found when they are stored for some period particularly in warm atmosphere. Alcoholic smell like wine is emitted from the fermented products. Such a condition results when sugar which is added to the product is hydrolyzed. A sugar molecule, *i.e.*, sucrose is hydrolyzed (= chemically broken down) to give rise to one molecule, each of glucose and fructose. These are hexose sugars and bring about fermentation. The condition usually results if too much citric acid is used in making the products. If such condition is observed, the jars containing them should be occasionally warmed as stated under the Heading 5 in this Chapter.

10. Microbial Spoilage

Jelly, jam and marmalade are prepared using high amount of sugar. Due to the presence of such sugar, micro-organisims usually do not infect them and if they remain present, they are killed by plasmolytic phenomenon. Thus, sugar acts as a preservative also apart from other two functions, *i.e.*, a dehydrating substance of pectin and a sweetening agent. Nevertheless, microbial growth is sometimes found in them. In such condition, the products should be taken out from the glass jars and heated by addition of some sugar and water. If microbial infection is much, it is advisable not to use the product and the jar containing the product should be thrown away.

11. Blackening Discoloration

Sometimes blackening appearance is found in the jelly, jam or marmalade on storing them. This is not very common and causes due to the presence of some chemicals that already remained present in the fruits. It is found that blackening discoloration of the products is usually increased when the jars containing them are maintained on direct exposure to bright light. Hence, if blackening is found to have started, the jars containing them should be transferred to a shady place.

Chapter 9

Method of Preparation of Artificial Jelly

Two steps are actually involved in the making of jelly from fruits (or other plant parts). These are firstly, to extract the pectin in boiling water and secondly, to add a dehydrating substance, *i.e.*, sugar in the pectin solution in water (= colloidal solution) to make formation of the jelly (= to set gel). Now, if already extracted pectin is used, obviously any fruit (or other plant part) would not be required to secure the pectin. This pectin could be made into solution in water and by addition of sugar, jelly could be prepared easily. This is the principle of making a jelly without use of any fruit or other plant part. When such a jelly is made, it is sold in the market as "artificial jelly" or "synthetic jelly."

An important aspect should not be overlooked in this context. That is, when a jelly is made by using fruits or other plant parts, not only pectin but also other compounds present in them which are soluble in water, *e.g.*, vitamins of the B group and C., mineral nutrients, amino acids, flavonoids, anthocyanins and such other compounds also come into the solution. These have tremendous nutritional importance. Thus, apart from pectin, these nutrients are present in the jelly when it is made from fruits or other plant parts. But these are not present in a jelly when it is made from extracted pectin, i.e., in an artificial jelly. Hence, jelly made from fruits or plant parts is nutritionally much superior than an artificial jelly.

Artificial, *i.e.*, synthetic jelly is prepared in the same way as jelly is made from fruit and as described in the Chapter 5. However, it can be made more easily and the time required would be much less. The salient steps in preparation are stated in the following:

1. Procurement of Pectin

Extracted pectin could be purchased from a reliable medical store. It is available in liquid or in powder form. Pectin is extracted by the chemical companies from wild apples, wild guavas, other fruits and peels of wild citrus fruits whose albedo portions are thick. On extraction, the pectin is purified in the chemical factories and several grades are made. Highly purified and ultra-pure grades are used in pharmaceutics and these are very costly. For the purpose of making jelly, ordinary or commercial grade pectin is suitable and also that this grade does not cost high.

2. Boiling in Water

The pectin should be taken in a stainless steel or aluminium pan to which water should be added and boiled. So much water as is used in making jelly from fruits is not necessary to use and a little quantity is considered satisfactory.

3. Addition of Sugar

When much of the water has evaporated, sugar should be added to it. To determine the amount of sugar to be added, the same method as described in Chapter 5 (6) should be adopted. To avoid burning of sugar, it should be added part by part with constant stirring to be done with a wooden ladle.

4. Addition of Artificial Colour and Flavour

Artificial jellies are sold in the market usually by the addition of various types of colouring (= dyeing) and flavouring chemicals. If it is desired to use them, they should in fact, be added towards the end, *i.e.*, about 5-10 minutes before the pan containing the boiling jelly is removed from fire.

A number of dyeing chemicals are used. A few are stated below:

To impart red colour, common dyes are Erythrosine, Allura red (syn. Red 40), Citrus red 2, Congo red (syn. Rodamin B) *etc.* For yellow colour, Trartrazine (syn. Yellow 5) and Sunset yellow (syn. Yellow 6) are commonly used. If green colour is needed, Fast green (syn. Green 3), Diamond green may be used. To impart blue colour, Brilliant blue (syn. Blue 1) or Indigo carmine may be used.

It should be well-remembered that all the dyeing chemicals are poisonous. They may cause allergy, hyperactivity in children, kidney tumour, brain tumour, enlarged bladder and even cancer *etc.* Therefore, it is not advisable to use them. If at all is used, none of these should be used beyond permissible limit.

Like dying chemicals, many manufacturing companies add flavouring chemicals to the jelly to impart flavor. Some of these chemicals are as follows:

To bring about ripe banana flavor, Iso-amyl acetate is used, for flavour of apple, Manzanate and for flavour of pear, Ethyl decadienoate are used. If flavour of citrus fruits is desired, Limonene and Octyl acetate may be used, for flavour of pineapple, Allyl hexanoate and to impart flavour of grape, Methyl anthranilate may be added. Flavouring chemicals are highly volatile. They should be added near the end of preparation and much heating after their addition will cause them to evaporate. In this connection, it is again brought to notice that colouring and flavouring chemicals are all harmful to health and their use above the permissible limit is never warrantable.

Rest of the procedure, *i.e.*, determining end-point, disinfection of glass-jars, filling them with jelly, solidification,covering with wax, labeling and storage should be done in the same way as have been described in the Chapter 5 (9-15). If an artificial jelly is made by adding flavour of any particular fruit, it should never be claimed that the jelly has been made from that fruit. Label of such a jelly should specify that the *jelly is artificial and has been prepared without use of any fruit* but flavour of that particular fruit has been added.

Chapter 10

Concluding Part

It has been mentioned earlier that nutritional, medicinal and other qualities of fresh fruits are of much greater importance than the products that are made from them. Hence, switching over fruits to make products, creating shortage of them as fresh fruits to the people is not warrantable although India's fruit production comes to about 99,000 thousand metric tonnes.

Making human consumable products from fruits has beyond doubt, its necessity under certain conditions. Apart from hotels, restaurants, voyages and use of tourists, dire necessity of them is felt for the defence personnels during war. To meet the regular demand of fruit products, utilization of those fruits which are less valued in their fresh from is a justifiable approach. An example may be cited as of guava. Quite a considerable part of guava produced in India is made use of in making jelly. It may be mentioned in this context that there are a number of wildly grown species of guava which are much suitable materials to make jelly as compared to the common guava fruit, *i.e., Psidium guajava* L. Some of these less known, wild species of guava may be named as *Psidium cattleianum* Sab., *P. guineense* Sw. (R.) (L.) Sk., *P. friedrichsthalianum* (O.B.) (Nie.) *etc.* Many of such less utilized, minor fruits that are found to grow in various parts of India as uncultivated may be suitably exploited in making jelly. Some examples may be cited as *Carissa carandas* L., swollen peduncle and thalamus of *Anacardium occientale* L., *Ziziphus mauritiana* lam., *Limonia acidissima* L., *Aegle marmelos* (L.) Corr. Ex. R., *Citrus mitis* Bl., *Physalis peruviana* L., *Averrhoa carambola* L., *Acca sellowiana* (O.B.) Bur. *etc.* High quality jelly is possible to be made from the peels of onion, mango and some other fruits and petals of roselle. The central core of jackfruit also makes jelly of good quality.

Fruit processing and preservation industry in commercial scale in India has started only 50 years before. Adequate studies aiming at improvement of preserved products are in infant stage now and need exploration. Blending jelly, jam, marmalade and other fruit products with spices and condiments has been an area that demands attention.

Apart from heavy industries, small-scale and cottage-industries to make fruit products need impetus. Some of the fruit products do not necessitate sophisticated equipments and could be prepared in small-scale units and with less capital.

Some of the fruit products, *e.g.*, beverages like mango nectar, guava nectar, clarified mango and banana juices *etc.* have great demand in the western countries. Pickeles have great demand in central Asian countries and to make these, fruits which have organoleptically less appeal are usually utilized. Higher production of fruit products of such type in the country needs emphasis as positive step to boost up industrial growth in the country.

Fruit products that are made in India are considered costly in the country and are difficult to be afforded by the economically less advanced people. A common cause of high price of the products is high price of cans, glass-jars, glass- bottles, packaging materials, transportation *etc.* Attention is necessary to explore less costly materials so that, the products reach the hands of consumers in the country at lower price.

Japanese plum: *Prunuts salicina* Lindl.

Karna: *Citus karna* Raf.

Karonda: *Carissa carandas* L.

Kumquat: *Fortunella* species

Lemon: *Citrus limon* Burma. And *Citrus medica* var. *limonum* (Lush.)

Litchi: *Litchi chinensis* Son.

Loquat: *Eriobotrya japonica* Lindl.

Mandarin orange: *Citrus reticulata* Blanco.

Mango: *Mangifera indica* L.

Mangosteen: *Garcinia mangostana* L.

Monkey jackfruit: *Artocarpus lacucha* Roxb.

Mulberry: *Morus* species

Otaheite gooseberry: *Phyllanthus acidus* Sk.

Papaya: *Carica papaya* L.

Peach: *Prunus persica* Bat.

Pear: *Pyrus communis* L.

Pecan nut: *Carya illinoensis* (Wang.) K. Koch

Phalsa: *Grewia subinaequalis* D.C.

Pineapple: *Ananas comosus* Merr.

Plantain: *Musa sanguine* Hook. f.

Plum: *Prunus domestica* L.

Poncirus: *Poncirus trifoliata* Raf.

Pomegranate: *Punica granatum* L.

Prune (dried alubokhara): *Prunus domestica* L.

Quince: *Cydonia oblonga* Mill.

Raspberry (orange colured): *Rubus ellipticus* Smith

Rose apple: *Syzygium jambos* Als.

Rough lemon (Jambhiri): *Citrus jambhiri* Lush.

Sapota: *Achras zapota* L.

Shaddock: *Citrus grandis* Osb.

Strawberry: *Fragaria* species and hybrids

Surinam cherry: *Eugenia uniflora* L.

Sweet orange: *Citrus sinensis* Osb.

Walnut: *Juglans regia* L.

Water-chestnut: *Trapa natans* L.

Water-melon: *Ctrullus lanatus* Th. (Mat. & Nak.)

Wax apple: *Syzygium samarangense* (Bl.) Mer. and Per.

Appendix II

Units and Conversion

Weight

1 kilogram	=	1000 gram = 0.1 quintal
	=	0.001 tonne = 2.205 lb
1 quintal	=	100 kilogram
1 tonne	=	10 quintal = 100 kilogram
1 gram	=	0.001 kilogram = 1000 milligram
	=	106 mcg = 109 nanogram
	=	0.353 ounce
1 mg	=	0.001 gram = 1000 microgram
1 microgram	=	0.001 milligram
1 pound	=	16 ounce = 453.6 gram
	=	128 dram (drachm)
1 ounce	=	28.35 gram = 8 dram
1 carat	=	0.2 gram = 200 milligram

Volume

1 litre	=	1000 ml = 0.001 cu.metre
	=	0.264 gallon = 1.76 pint
1 millilitre	=	0.001 l = 1000 microlitre
	=	1000 nanolitre
1 microlitre	=	0.001 millilitre
1 fl. dr	=	60 minims
1 fl. ounce	=	8 fl. dram
1 pint	=	0.57 litre = 20 fluid ounce
1 gallon	=	8 pint = 3.785 litre

(1 mouthful of liquid in an adult male : 110-120 ml and in an adult female person : 100-110 ml)

Length

1 metre	=	100 cm = 1000 mm
	=	3.281 feet = 39.37 inch
1 kilometre	=	1000 metre = 1093.6 yard
	=	0.62 mile
1 centimetre	=	0.39 inch
1 mile	=	1.61 km = 1760 yard
	=	8 furlong = 5280 feet
1 foot	=	0.3048 metre
1 yard	=	3 feet = 36 inch
	=	0.9144 metre
1 inch	=	2.54 centimetre

Index

G

H

I

J

L

M

O

P

S